SOIL HUMIC SUBSTANCES:
ELECTRON TRANSFER PROCESS AND ENVIRONMENTAL IMPACTS

土壤腐殖质
电子转移过程及其环境影响

余 红 檀文炳 席北斗 等著

化学工业出版社

·北京·

内容简介

本书以土壤腐殖质作为研究对象，评估土壤原位固相腐殖质的电子转移能力，明确影响土壤原位固相腐殖质电子转移能力的关键性因素，揭示土壤腐殖质在生物地球化学的氧化还原过程中作为胞外电子穿梭体的持续能力，阐明土壤腐殖质电子转移能力对气候变暖和土地利用的响应机制，为研究土壤污染物的转化与积累机制提供一定的理论依据，而且在土壤污染修复方面具有重要的实践意义。

本书具有较强的知识性和针对性，可供从事土壤结构、功能及污染防控等领域的工程技术人员和科研人员参考，也可供高等学校环境工程、生物工程和化学工程及相关专业师生参阅。

图书在版编目（CIP）数据

土壤腐殖质电子转移过程及其环境影响/余红等著. —北京：化学工业出版社，2020.12
ISBN 978-7-122-37879-8

I.①土… II.①余… III.①土壤-腐殖质-研究 IV.①S153.6

中国版本图书馆 CIP 数据核字（2020）第 192981 号

责任编辑：刘兰妹　刘兴春　　　　　　　文字编辑：李　玥
责任校对：李雨晴　　　　　　　　　　　装帧设计：张　辉

出版发行：化学工业出版社（北京市东城区青年湖南街 13 号　邮政编码 100011）
印　　装：涿州市般润文化传播有限公司
710mm×1000mm　1/16　印张 12　彩插 8　字数 210 千字　2021 年 4 月北京第 1 版第 1 次印刷

购书咨询：010-64518888　　　　　　　　售后服务：010-64518899
网　　址：http://www.cip.com.cn
凡购买本书，如有缺损质量问题，本社销售中心负责调换。

定　　价：98.00 元　　　　　　　　　　　　　　　　版权所有　违者必究

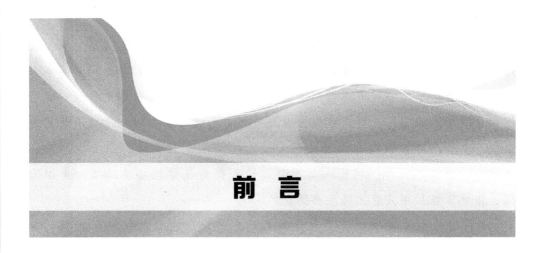

前 言

土壤有机质是指土壤中动物、植物与微生物残体的不同分解、合成阶段的各种含碳有机化合物。土壤有机质是土壤的重要组成成分，是保障土壤具有结构性与生物性的重要条件，是与生命活动密切相关的基本物质。尽管土壤有机质的含量仅占土壤总量的很小一部分，但它对土壤形成、土壤肥力、土壤生物地球化学过程、环境保护及农林业可持续发展等方面都有着极其重要的意义。正因为土壤有机质如此重要，目前关于土壤有机质的研究已经有200多年的历史，自21世纪以来被称为地球环境领域里最后的前沿。

土壤腐殖质是指新鲜有机质经过微生物分解转化所形成的一类大分子有机化合物，是土壤有机质的主要组成部分。在还原条件下，溶解性的和颗粒态的腐殖质既可作为电子受体，接受从微生物传递出的电子，同时还原后的腐殖质又可作为电子供体，将电子传递给铁氧化物等矿物以及各种各样的有机污染物和无机污染物，而且其在接受电子或供给电子的过程中具有一定的可逆性，从而使其在许多生物地球化学的氧化还原过程中可以充当胞外电子穿梭体的角色，这在很大程度上会对具备氧化还原活性的有机污染物和无机污染物的氧化还原动力学特征与降解途径产生深刻影响。

本书总结了笔者及其团队多年的研究工作，以土壤腐殖质作为研究对象，评估土壤原位固相腐殖质的电子转移能力，明确影响土壤原位固相腐殖质电子转移能力的关键性因素，揭示土壤腐殖质在生物地球化学的氧化还原过程中作为胞外电子穿梭体的持续能力，阐明土壤腐殖质电子转移能力对气候变暖和土地利用的响应机制。本书共6章，第1章为绪论，主要介绍了土壤腐殖质电子转移过程的国内外研究进展；第 2 章为土壤原位固相腐殖质的电子转移能力；第 3 章为土壤溶解性腐殖质电子循环能力；第 4 章为土壤腐殖质电子转移能力对增温的响应；第 5 章为土壤腐殖质竞争性抑制甲烷生成对增温的响应；第 6 章为土壤腐殖质电子转移能力对土地利用变化的响应。全书具有较强的知识性和针对性，可供从事土壤结构、功能及污染防控的工程技术人员、科研人员参考，也可供高等学校环境工程、生物工程、化学工程及相关专业师生参阅。

本书由余红、檀文炳、席北斗等著，全书最后由余红统稿并定稿。本书的出版获得了生态环境部综合司 2020 年城市环境规划管理项目经费资助。此外，本书在编写过程中参考了部分相关领域的文献，引用了国内外许多专家和学者的成果和图表资料，谨此向有关作者致以谢忱。

鉴于笔者编写时间和学术水平有限，书中存在不足和疏漏之处在所难免，敬请专家、学者及广大读者批评指正。

著者

2020 年 8 月

目 录

第1章 绪论 ··· 1

1.1 研究背景 ··· 1

1.2 国内外研究现状及发展趋势 ··· 2

   1.2.1 生物地球氧化还原——全球物质循环的驱动力 ····················· 2

   1.2.2 微生物胞外呼吸——基于界面电子转移的氧化还原过程 ············· 8

   1.2.3 电子穿梭体——微生物胞外呼吸的"催化剂" ····················· 28

   1.2.4 土壤腐殖质——电子穿梭体的"聚合体" ························· 32

1.3 研究思路与研究内容 ·· 40

   1.3.1 研究目标 ··· 40

   1.3.2 研究内容 ··· 40

参考文献 ··· 41

第2章 土壤原位固相腐殖质电子转移能力 ································· 64

2.1 土壤原位固相腐殖质电子转移能力——Fe（Ⅲ）还原的证据 ············· 66

   2.1.1 土壤固相腐殖质的微生物还原反应中的胞外电子转移过程 ········· 66

   2.1.2 土壤腐殖质电子转移能力及其与物理化学性质之间的关系 ········· 67

   2.1.3 土壤固相腐殖质的微生物可利用性对其电子穿梭能力的影响 ······· 76

   2.1.4 不同土壤团聚体组分中固相腐殖质的电子转移过程的异质性 ······· 79

2.2 土壤原位固相腐殖质电子转移能力——电化学测量的证据 ·················· 82
  2.2.1 微生物还原前后土壤中固相腐殖质的氧化还原特性 ················· 82
  2.2.2 不同土壤团聚体中固相腐殖质的氧化还原特性 ···················· 86
  2.2.3 环境意义 ··· 88
参考文献 ·· 88

第3章 土壤溶解性腐殖质电子循环能力 ·· 93

3.1 厌氧条件下微生物还原后腐殖质的还原能力 ······························· 94
3.2 氧气再氧化后微生物还原腐殖质的还原能力 ······························ 102
3.3 微生物还原和氧气再氧化循环过程中腐殖质的电子循环能力 ········· 113
3.4 胡敏酸与富里酸之间氧化还原循环能力的比较 ·························· 115
3.5 环境意义 ··· 116
参考文献 ·· 117

第4章 土壤腐殖质电子转移能力对增温的响应 ································ 121

4.1 土壤腐殖质电子转移能力与温度的关系 ··································· 122
4.2 土壤腐殖质电子转移能力与其化学结构的关系 ·························· 130
4.3 土壤和植物凋落物对土壤腐殖质化学结构的影响 ······················· 136
4.4 土壤和植物凋落物与温度的关系 ··· 143
参考文献 ·· 145

第5章 土壤腐殖质竞争性抑制甲烷生成对增温的响应 ······················· 151

5.1 添加腐殖质对厌氧条件下甲烷和二氧化碳生成的影响 ················· 152
5.2 腐殖质微生物还原反应竞争性抑制甲烷生成对温度升高的响应 ······ 154
5.3 环境意义 ··· 156
参考文献 ·· 157

第6章 土壤腐殖质电子转移能力对土地利用变化的响应 ···················· 160

6.1 土壤腐殖质电子转移能力对农用地类型的响应 ·························· 160

  6.1.1 不同农用地类型下的土壤腐殖质电子转移能力 …………………… 161

  6.1.2 土壤腐殖质化学结构对电子转移能力的影响 …………………… 161

  6.1.3 土壤腐殖质转化和分解对其化学结构的影响 …………………… 165

  6.1.4 环境意义 …………………………………………………………… 167

6.2 土壤腐殖质电子转移能力对水稻田耕作年限的响应 ………………………… 169

  6.2.1 原位固相天然有机质对微生物还原 Fe(Ⅲ) 矿物的影响 ………… 170

  6.2.2 微生物种类对微生物还原 Fe(Ⅲ) 矿物的影响 …………………… 173

  6.2.3 基于固相 Fe(Ⅲ) 矿物的腐殖质电子转移能力对水稻田耕作

     年限的响应 ………………………………………………………… 173

参考文献 ………………………………………………………………………………… 178

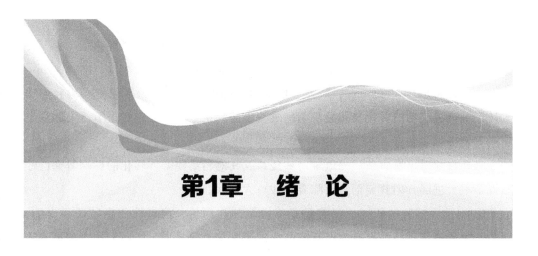

第1章 绪 论

1.1 研究背景

土壤是自然环境重要的组成部分,是人类赖以生存的、最重要的可再生自然资源和永恒的生产资料,土壤环境质量对人类生命健康与安全和整个社会的稳定与发展具有战略性意义。我国的土壤污染与防治的情况不容乐观,重金属污染土壤、农药化肥的不合理使用导致污染土壤的面积超过千万公顷,这不仅直接关系到农产品的安全和质量问题,而且严重阻碍了社会可持续发展。因此,土壤迫切需要保护,进行土壤污染防治与修复已经成为全球范围内许多国家共同关注的问题,是政治、商业和学术领域讨论的热点。

氧化还原反应是陆地生态系统中普遍存在的且重要的化学或生物化学反应(Fiedler et al.,2007),其本质是通过电子转移来实现的。微生物胞外电子传递过程是新发现的新型微生物厌氧能量代谢方式,它作为一种独特的氧化还原过程,在污染物原位修复、污水处理以及清洁生物能源提取方面已经逐渐呈现出不可替代的优越性和重要的应用前景(Jiang et al.,2009;Marsili et al.,2008;Rosenbaum et al.,2012;White et al.,2013;Kappler et al.,2014),因此近几年来受到科学界的广泛关注。

腐殖质(humic substances,HS)是一类具有氧化还原性质的天然有机化合物,它在土壤中具有广泛的分布。在还原条件下,溶解性腐殖质既可作为电子受体,接受从微生物传递出的电子(Lovley et al.,1996),同时还原后的腐殖质又可作为电子供体,将电子传递给铁氧化物等矿物(Piepenbrock et al.,2014),而且其在接受电子或供给电子的过程中具有一定的可逆性,从而使其在许多生物地球化学的氧化还原过程中可以充当胞外电子穿梭体的角色,这在很大程度上会对具备有氧化还原活性的有机污染物和无机污染物的氧化还原动力学与降解途径产生重要影响（Borch et al.，2005；Fulda et al.，2013；Maurer et al.，2012，2013；Lovley et al.，2000）。由此可见，研究土壤腐殖质的电子转移能力不仅可以为我们深入认识土壤污染物的转化与积累机制提供一定的理论依据，而且在土壤污染修复方面具有重要的实践意义。

尽管目前对土壤腐殖质的电子转移规律及其对污染物降解转化的影响研究已经取得了很大进展，但我们仍然面临着许多重大挑战。

① 目前大部分的研究成果都是基于提取纯化后的、溶解性的腐殖质得出的，而关于土壤原位固相腐殖质电子转移能力的研究却鲜有报道。实际上，土壤提取纯化后的溶解性腐殖质并不能等同于土壤原位固相腐殖质，因为土壤原位固相腐殖质的许多理化特性和所处的微环境是提取纯化后的溶解性腐殖质所没有的（Kelleher and Simpson，2006；Lehmann et al.，2008；Schmidt et al.，2011）。

② 土壤是一个开放的体系，土壤腐殖质处于不断更新之中，这对于土壤整体而言，它有持续的电子穿梭体功能，但对于土壤腐殖质本身而言，间歇性缺氧-曝气的土壤环境会对腐殖质的末端电子穿梭功能产生竞争性的促进或抑制作用（Klüpfel et al.，2014），使得土壤环境中腐殖质电子穿梭体功能并非是持续恒定的，而是有一定的"寿命"的，因此土壤腐殖质电子穿梭体功能的持续能力也是一个有待进一步研究的重要问题。

③ 人类活动导致的气候变暖以及土地利用变化是当前存在的重要环境问题，它们可以直接地或间接地对土壤腐殖质的形成、分子结构、理化性质以及与矿物的相互作用产生重大影响（Han et al.，2011；Becerra-Castro et al.，2015），这在一定程度上会进一步造成土壤腐殖质的电子转移能力发生相应变化。

然而，目前关于气候变暖和土地利用变化究竟是如何影响土壤腐殖质的电子转移能力还不清楚。显而易见，弄清这种影响机理对于在气候变暖和土地利用变化背景下如何更好管理土壤和维持土壤健康具有重要指导意义。

1.2 国内外研究现状及发展趋势

1.2.1 生物地球氧化还原——全球物质循环的驱动力

许多元素如碳（C）、氮（N）、硫（S）、铁（Fe）和锰（Mn）以及易于氧化还原的微量元素如砷（As）、铬（Cr）、铜（Cu）、铀（U）的生物地球化学循环是由氧化还原过程驱使的。这些元素的化学形态、生物可利用性、毒性、在环境中的移动性直接受氧化还原反应的影响。此外，其他不具有氧化还原活性的元素和化合物的生物地球化学行为可能间接地与天然有机物和矿物的氧化还原转换有关，特别是铁的氢氧化物、锰矿物、含铁黏土矿物、铁硫化物。氧化还原活性基团与腐殖质和矿物表面相结合可进一步催化离子和分子，包括许多有机污染物的氧化或还原（Ginder-Vogel et al.，2005；Kappler and Haderlein，2003；Polizzotto et al.，2008；Wu et al.，2006）。由此可见，氧化还原过程可以为环境污染治理工程提供新的战略机遇，了解环境中氧化还原界面的生物地球化学过程对保护环境生态系统健康是至关重要的。

1.2.1.1 氧化还原过程与碳、氮、磷元素循环

地球上所有生命的能量都来源于氧化还原过程。生物质的生产需要通过电子转移将碳元素、常量营养元素（如 N、S）和微量营养元素（如 Fe、Mn）形成合适的氧化态进入生物分子中。地球诞生之前，生物氧化还原活性已经使氧化性的环境表面覆盖了还原性物质，如有机物、硫化物和甲烷。沿着全球性的氧化还原梯度，许多潜在的电子供体、电子受体和碳源可以合成大量的在生态和代谢上具有多样性的微生物（DeLong and Pace，2001）。

碳循环是由有氧光合作用驱动的，它通过失去水中的电子，固定二氧化碳，并产生氧气。另外，非氧光合作用和化学无机自营养所固定的碳可能是局域性环境尤其是极端环境下最重要的生物质来源（D'Hondt et al.，2002）。有机质和其他还原性物质前期在沉积物中的埋藏以及后期隆起暴露在地球表面所发生的风化氧化过程是控制地质时间尺度上的大气组成和地球气候变化的关键过程（Berner，1999）。

环境中氮、磷的行为与碳的氧化还原过程密切相关。氮存在多种氧化态，它们的许多氧化还原过程，如固氮、硝化、反硝化、异化硝酸盐还原为氨都是由微生物驱动的。这些微生物过程影响氮素的供应，进而影响当地乃至全球范围内有机质的生产和循环（Howarth，2002）。类似这样的关联同样存在于铁、硫、磷和重金属元素的氧化还原循环中（Moore，2014）。例如，黄铁矿氧化作用是含水层中硝酸盐循环的重要途径，然而，该过程产生的硫酸盐可能刺激硫酸盐的微生物还原，这反过来可能会导致硫酸盐将溶解态氢氧化铁（Ⅲ）还原（Flynn et al.，2014；Friedrich and Finster，2014）。再有，当富含磷酸盐的地下水排放进入地表水体可能会造成富营养化（Lucassen et al.，2004）。上述这些例子说明了生物地球化学氧化还原循环之间存在高度的耦合。

1.2.1.2 氧化还原过程与铁、锰矿物的动态变化

作为地球表面最丰富的过渡金属，铁在环境生物地球化学中发挥着重要作用。氧化态铁在极端酸性条件下可溶，但在近中性 pH 条件下 Fe(Ⅲ) 沉淀形成氢氧化铁，氢氧化铁表面能够催化环境中许多重要的氧化还原反应。在还原条件下，氢氧化铁会被无机物如硫化物还原，从而可能释放出有害的吸附物（Afonso and Stumm，1992）。此外，Fe(Ⅲ) 矿物可能作为终端电子受体被异化铁还原细菌微生物还原。这些微生物在土壤、淡水和海洋环境中无处不在，它们能够将有机化合物在细胞质内的氧化和细胞外难溶 Fe(Ⅲ) 矿物质的还原结合起来，从而通过磷酸化作用中的电子转移获得能量（DiChristina et al.，2005）。Fe(Ⅲ) 矿物的还原产生可溶性 Fe(Ⅱ) 和各种次生矿物，包括 Fe(Ⅱ) 矿，如蓝铁矿 [$Fe_3(PO_4)_2$] 和菱铁矿（$FeCO_3$），Fe(Ⅲ) 矿物（如针铁矿）和 Fe(Ⅱ、Ⅲ) 混合铁矿，如磁铁矿（Fe_3O_4）和绿铁锈（层状双氢氧化物）。溶解态、吸附态和固态 Fe(Ⅱ) 在一系列生物氧化还原过程中可以作为强还原剂（Liger et al.，1999；Lloyd et al.，2000）。

Fe(Ⅱ) 的氧化可由好氧微生物和厌氧微生物介导。微生物介导的 Fe(Ⅱ) 氧化在较低 pH 值环境下就可以进行，但在这样的条件下化学氧化则无法进行。在近中性 pH 条件下，Fe(Ⅱ) 的微生物氧化将会和 Fe(Ⅱ) 的快速化学氧化竞争。因此，在氧含量低的条件下特别是好氧-厌氧界面，例如被水淹没的植物根部附近，这些氧化铁(Ⅱ) 的微生物将占据一定的优势（Druschel et al.，2008）。在嗜中性、缺氧的环境中，硝酸盐依赖型 Fe(Ⅱ) 氧化微生物可利用硝酸盐和氧化锰(Ⅳ) 作为电子穿梭体来促使 Fe(Ⅱ) 的氧化（Kappler and Straub，2005）。

锰氧化物矿物会参与环境中各种氧化还原反应，例如，水钠锰矿（$\delta\text{-}MnO_2$）直接氧化硒(Ⅳ) 为硒(Ⅵ)，氧化铬(Ⅲ) 为铬(Ⅵ)，氧化砷(Ⅲ) 为砷(Ⅴ)（Post，1999）。锰(Ⅱ) 的氧化可发生在各种环境中并可由多种细菌和真菌介导（Miyata et al.，2006）。生物介导的锰(Ⅱ) 氧化所产生的最初产物通常结晶性很差，为层状氧化锰(Ⅳ) 矿物。虽然锰(Ⅱ) 氧化的最终矿物形态往往取决于锰(Ⅱ) 氧化期间及之后的地球化学条件，但生物介导的锰(Ⅱ) 氧化通常被认为是环境中锰氧化物的主要来源途径（Tebo et al.，2004）。

自然界水体中铁(Ⅲ) 和锰(Ⅲ、Ⅵ) 矿表面结构和反应活性会受无机吸附物和天然有机物影响（Bauer and Kappler，2009；Borch et al.，2007）。腐殖质具有氧化还原活性，其作为微生物与矿物质之间的电子穿梭体能够促进不溶性 Fe(Ⅲ) 矿物为微生物利用（Kappler et al.，2004）。腐殖质还可以通过络合、增加金属离

子可溶性、在矿物表面吸附等方式进一步影响铁、锰矿物的形成，进而导致结晶矿石无法形成（Eusterhues et al., 2008; Jones et al., 2009）。$Fe(Ⅲ)$ 和 $Mn(Ⅳ)$ 对天然有机物、磷酸盐、碳酸盐的强烈吸附可能会导致其所吸附的污染物发生解吸，但也可以阻止矿物对微生物的吸附，从而保护固相物质不被酶还原（Borch et al., 2007）。

1.2.1.3 氧化还原过程与重金属转化

一些微量重金属，如铬、铜、钴、银、锝和汞可以一些氧化态存在，其还原转化可能以化学方式进行，如铜（Ⅱ）可被二价铁离子（Fe^{2+}）或者硫氢酸（H_2S）还原为 $Cu(Ⅰ)$，$Cu(Ⅱ)$、$Ag(Ⅱ)$ 和 $Hg(Ⅱ)$ 可被由绿锈中存在的 $Fe(Ⅱ)$ 还原为基本态（O'Loughlin et al., 2003）。此外，微生物也可以通过异化或者解毒途径直接还原具有一定毒性甚至剧毒的金属（如 Cr、Hg、U）(Lovley, 1993)。例如，最近基于 X 射线吸收近边结构（XANES）和透射电镜（TEM）的研究结果表明，洪涝时岸边的污染土壤中金属铜的形成可能是一种细菌解毒的结果（Weber et al., 2009a; Weber et al., 2009b）。微量金属的还原可以降低其移动性，比如可溶性铬（Ⅵ）还原为难溶性铬（Ⅲ），$Hg(Ⅱ)$ 还原为挥发性 $Hg(0)$（Lovley, 1993）。

微生物硫酸盐还原作用可能导致难溶性亲铜重金属元素的沉淀（Kirk, 2004）。在富含硫的沉积物中，亲铜重金属元素可能与铁硫化物发生共沉淀或者形成其他金属的硫化物（Morse and Luther, 1999）。然而，在被污染的淡水湿地中发现亲铜重金属元素的可利用性可能超过可还原硫化物，亲铜重金属元素的动态变化可能受生物硫化物共沉淀的竞争影响（Weber et al., 2009b）。与难溶金属硫化物的共沉淀作用相比，金属硫化物族的形成可能更会显著提高金属在厌氧环境中的流动性，而且由于其在动力学上的稳定性，使其可以在好氧水体中存在（Luther and Rickard, 2005）。最近发现，在污染的河岸土壤中硫酸盐的还原作用可通过促进富铜硫化物胶体的形成来提高铜、铅和镉的移动性（Weber et al., 2009a），这与金属硫化物胶体能够提高水体等其他环境介质中污染物的移动性的说法相吻合（Deonarine and Hsu-Kim, 2009）。

砷的移动性、生物可利用性、毒性和环境命运受生物地球化学过程所控制，其控制方式主要是通过形成或者破坏砷的承载相、改变砷的氧化还原状态与改变砷的化学组成（Dixit and Hering, 2003）。在中性 pH 条件下，$Fe(Ⅱ)$ 的非生物和微生物氧化使之转化为溶解性较小的 $Fe(Ⅲ)$ 矿物，从而导致砷被吸附在次生铁矿物中而被固定（Dixit and Hering，2003）。然而，厌氧光合细菌、硝酸还原菌、Fe(Ⅱ)氧化细菌也能耐受高浓度砷并提供合适的途径来使得Fe(Ⅲ)与砷发生共沉淀（Hohmann et al.，2010）。在富含氢氧化物铁（Ⅲ）的蓄水沉积物中，持续的、长期的还原条件可能会使得吸附剂耗尽，从而发生砷的移动（Tufano and Fendorf，2008），但这种现象是极其罕见的。总之，Fe(Ⅲ)和Fe(Ⅱ)直接的生物地球化学氧化还原过程能够控制砷的移动性。

砷的移动性和毒性不仅受存在的合适的吸附剂的影响，同时也受其氧化还原形态的影响，通常认为砷还原态比其氧化态更易移动、具有更高毒性。许多生物地球化学过程可以直接或间接导致砷的氧化还原，例如Fe(Ⅱ)-针铁矿系统，或过氧化氢反应形成的Fe(Ⅳ)可以氧化砷(Ⅲ)(Amstaetter et al.，2010；Hug and Leupin，2003)。还有学者提出细菌也可以通过将As(Ⅴ)还原或者As(Ⅲ)氧化而改变其氧化还原状态来控制砷的移动性和毒性（Tufano et al.，2008），尽管这还不具有普遍性，但其反映出微生物对砷具有一定的解毒功能（Kulp et al.，2008）。此外，有些微生物还可以分泌活性有机或无机化合物与As(Ⅲ)或As(Ⅴ)发生氧化还原反应。最近研究还表明，腐殖质和醌模型化合物中的半醌自由基和氢醌也可以氧化砷（Ⅲ）为As(Ⅴ)，或还原As(Ⅴ)为As(Ⅲ)(Jiang et al.，2009；Redman et al.，2002)。

在好氧环境中，铀普遍以六价氧化态铀[UO_2^{2+}]存在，U(Ⅵ)在大多数环境条件下是可溶的，尤其是当U(Ⅵ)与碳酸盐结合时溶解性更大（Guillaumont et al.，2003）。相反，U(Ⅳ)溶解性较小，即使有地下水配体如碳酸钙存在时，U(Ⅳ)仍趋于相对稳定（Wu et al.，2006；Wu et al.，2007）。事实上，目前正在探索一种可能的铀微生物修复技术，实现原位修复U(Ⅵ)为U(Ⅳ)(Williams et al.，2013)。

在低温地球化学环境中，非生物U(Ⅵ)的还原可通过若干途径进行，但这些途径会受到许多条件的限制（Borch et al.，2010）。相反，许多普通的金属还原菌和硫酸盐还原菌可以将有机质与H_2的氧化作用和铀（Ⅵ）的还原作用联系起来，从而生成铀（Ⅳ）和晶体铀矿沉淀（Gorby and Lovley，1992）。然而，作为电子受体的硝酸盐或氢氧化铁（Ⅲ）的存在和潜在的U(Ⅳ)氧化剂可能在一定程度上会阻碍U(Ⅵ)的生物还原（Stewart et al.，2009）。

潜在的UO_2氧化剂包括分子氧、硝酸盐、硝酸盐还原中间体、氢氧化锰（Ⅳ）和氢氧化铁（Ⅲ）(Borch et al.，2010)。此外，UO_2还可以被生物活性物质催化氧化。硝酸盐是一种常见的铀的共污染物质，它不仅能够阻碍铀的生物还原，还有可能氧化铀（Ⅳ）（Senko et al.，2005）。硝酸盐氧化铀在热力学上是可行的，但在动力学上是受限的（Senko et al.，2005）。尽管由生物介导的将 NO_3^- 转化为 NO_2^-、NO 和 N_2O 可以提高 U(Ⅳ) 的氧化速率，但与 Fe(Ⅲ) 和 O_2 氧化 U(Ⅳ) 的速率相比还是显得相当缓慢。在普通地下水环境中，U(Ⅳ)/U(Ⅵ) 的氧化还原电位和 Fe(Ⅲ) 氢氧化物/Fe(Ⅱ) 的氧化还原电位相类似，因此，液相和固相的化学物质发生微小的变化都会导致 Fe(Ⅲ) 氢氧化物氧化 UO_2 的途径在热力学上时而可行、时而不可行（Ginder-Vogel et al.，2006）。脱氮硫杆菌和地杆菌能够催化硝酸盐而促进 U(Ⅳ) 的氧化，但目前尚不清楚这两种细菌是否能够从这个过程中获得能量（Beller，2005）。

相对于 U，Pu 和 Tc 的生物地球化学循环的特点不明显。在环境系统中 Tc 主要以可溶性 Tc(Ⅶ) 或相对可移动性 Tc(Ⅳ) 存在。在缺氧条件下，许多细菌能够催化还原 Tc(Ⅶ) 为 Tc(Ⅳ)，形成 Tc(Ⅳ) 氧化物（Fredrickson et al.，2004）。此外，溶解的和吸附的 Fe(Ⅱ) 作为一种微生物的还原产物也是 Tc(Ⅶ) 的强还原剂，然而氧气又会将 Tc(Ⅳ) 氧化可能是阻碍还原态锝固定的一个重大因素（Fredrickson et al.，2004）。

在环境中，Pu 通常以 Pu(Ⅲ)、Pu(Ⅳ)、Pu(Ⅴ) 或 Pu(Ⅵ) 态形式存在。通常认为 Pu(Ⅳ) 移动性较差，而 Pu(Ⅵ) 移动性最强，而 Pu(Ⅴ) 通常只是在环境中短暂存在（Icopini et al.，2009）。Pu(Ⅲ)、Pu(Ⅳ)、Pu(Ⅴ) 通常会被 Mn(Ⅳ) 氧化成 Pu(Ⅵ) 从而广泛存在于环境中（Powell et al.，2006）。与 U 相似，氧化态 Pu(Ⅴ、Ⅵ) 也可以被金属还原细菌如地杆菌和希瓦氏菌还原为 Pu(Ⅳ) 或 Pu(Ⅲ)（Icopini et al.，2009）。

1.2.1.4 氧化还原过程与有机污染物降解

有机污染物在环境中广泛分布，锰、铁等矿物质以及腐殖质可能会大大影响到众多有机污染物的氧化和还原转化过程。这些反应途径与速率取决于矿物类型、化学溶液和微生物活性。同样该过程也可以设计成修复策略，如通过可渗透性反应层（PRB）来减小地下水中有机（和无机）污染物的蔓延。

几项研究结果已经表明，相对于纯净的 Fe^{2+}，Fe(Ⅲ) 氧化态，如与 Fe(Ⅱ) 反应的赤铁矿可以显著提高许多还原性污染物如硝基芳烃、氯化溶剂、农药和消毒剂的转化速度（Amonette et al.，2000；Borch et al.，2005；Hakala et al.，2007），但还没有研究清楚的是与 Fe(Ⅱ) 反应为何会提高 Fe(Ⅲ) 氧化物的表面活性。Vikesland 和 Valentine（2002）研究了各种铁氧化表面 Fe(Ⅱ) 和氯胺反应的动力学，发现铁的氧化物对这些反应起着至关重要的作用。此外，针铁矿活化的Fe(Ⅱ)还原硝基苯反应只能发生在Fe(Ⅱ)水溶液中（Williams and Scherer，2004）。Fe(Ⅱ)氧化还原活性的增强不仅在氧化物存在时能观察到，在其他主要铁矿物存在时也能观察到。六氯乙烷和4-氯硝基苯的表面反应速率顺序为：Fe(Ⅱ)＋菱铁矿＜Fe(Ⅱ)＋铁氧化物＜Fe(Ⅱ)＋铁硫化物（Elsner et al.，2004）。磁铁矿的粒径大小和聚集状态会影响四氯甲烷的还原转化（Vikesland et al.，2007）。另外，最近的一项研究表明，磁铁矿中Fe(Ⅱ)/Fe(Ⅲ)的化学计量比可能会改变磁铁矿颗粒的氧化还原性质，从而更有利于硝基苯还原（Gorski and Scherer，2009）。

细菌与厌氧性有机污染物如甲苯和氯乙烯相结合能够还原电子载体（如腐殖质），被还原的电子载体可以进一步转移电子到一些吸电子化合物中，如偶氮染料、多卤代化合物、硝基芳烃（Van der Zee and Cervantes，2009）。研究显示，氢醌基团还原的胡敏酸和模型化合物（如AQDS）会对硝基化合物的还原动力学和降解途径都产生显著影响（Borch et al.，2005；Van der Zee and Cervantes，2009）。

除了通过金属氧化物和有机物质在自然条件下转化污染物，PRB代表了处理污染地下水的一种环境修复技术（Gillham and Ohannesin，1994）。以零价铁为基础的PRB已经被证明可以有效去除包括卤化有机溶剂在内的多种污染物（O'Hannesin and Gillham，1998）。通常情况下，随着时间的推移，零价铁被氧化形成氢氧化铁导致PRB反应活性降低。然而，最近的研究表明，通过铁还原细菌的生物强化作用有可能改善PRB性能（Van Nooten et al.，2008）。

相对于零价铁对有机污染物的还原处理，高锰酸盐是通常用于有机化合物的原位化学氧化剂，其有效pH值范围宽，易于处理，而成本相对较低。高锰酸钾能够氧化多种有机污染物，如1,4-二氧杂环己烷、甲基叔丁基二甲醚、甲基乙基酮、爆炸物［如三硝基甲苯（TNT）］、农药（如涕灭威和敌敌畏）、酚类化合物（如对硝基苯酚）和氯化物（如四氯乙烯）(Waldemer and Tratnyek，2005）。另外，最近的研究还表明，土壤中常见的锰氧化物（如水钠锰矿）也可以氧化常见的污染物如抗菌药物（如酚、氟喹诺酮类、芳香性的N-氧化物和四环素类）、双酚A以及17α-炔雌醇（Lin et al.，2009；Zhang et al.，2008）。

1.2.2 微生物胞外呼吸——基于界面电子转移的氧化还原过程

微生物胞外呼吸（extracellular respiration）是近年来新发现的在厌氧条件下微生物能量代谢的方式，指厌氧条件下微生物在胞内彻底氧化有机物释放电子，产生的电子经胞内呼吸链传递到胞外电子受体使其还原，并产生能量维持微生物自身生长的过程。它是一种涉及电子在微生物细胞与胞外电子受体/电子供体之间传递的呼吸方式。从化学角度来看，微生物胞外呼吸本质上是一种由微生物介导的氧化还原过程。在理论方面，微生物胞外呼吸的发现为呼吸链电子传递、胞外电子转移、能量产生途径等科学问题提供了新的视角。在应用方面，微生物胞外呼吸在碳、氮、硫等元素生物地球化学循环、污染物转化消减和微生物产电等方面发挥了积极作用（Lovley et al., 2004），表现出巨大的应用潜力。

1.2.2.1 微生物胞外呼吸菌

微生物胞外呼吸菌在环境中广泛存在，人们已经在土壤、泥炭、污泥、湖泊沉积物、河流沉积物、海洋沉积物以及水体等环境介质中分离富集出了许多具有胞外呼吸功能的微生物。根据胞外电子受体的不同，微生物胞外呼吸菌主要分为腐殖质还原菌、异化金属还原菌和产电微生物。除了常规微生物，许多极端环境微生物也具有胞外电子传递能力，如嗜热菌、嗜酸菌、嗜碱菌等。按照对氧气的需求，胞外呼吸菌又分为兼性厌氧菌和严格厌氧菌。胞外呼吸菌大部分集中在变形杆菌门（Proteobacteria）、放线菌门（Acidobacteria）与厚壁菌门（Firmicutes）三个门。已发现的胞外呼吸菌种大多数为革兰氏阴性菌，只有少数为阳性菌。产电微生物主要是人工驯化产生的，通过对各种沉积物进行电极富集培养实验发现，大部分产电微生物菌种的16S rRNA基因序列与δ-变形菌（*Desulfoarculus baarsii*）的相似度可达76%~95%（Holmes et al., 2004）。目前报道的胞外呼吸菌的数量仅占自然界的极小部分，而且很多菌的功能机制还不完全清楚。随着研究的不断深入以及微生物分离方法和分子生物学方法的不断完善，胞外呼吸菌资源将会持续不断被发现和丰富。

有些微生物有多种胞外呼吸途径的功能，例如希瓦氏菌属（*Shewanella*）和地杆菌属（*Geobacter*）。这两种菌属也是目前研究最深入和最系统的胞外呼吸菌，已经成为异化金属还原菌群中的模式菌（Weber et al., 2006），而且希瓦氏菌（*Shewanella oneidensis* MR-1）和地杆菌（*Geobacter sulfurreducens* PCA）的基因组已经全部测序完成（Heidelberg et al., 2002；Methé et al., 2003）。两种微生物基因组的研究发现，在希瓦氏菌和地杆菌中分别有42个和111个与细胞膜相关的细胞色素c基因，但已经确定功能的细胞色素c只有几种，大量的基因功能还不完全清楚，需要进一步探索。正是由于这些大量蛋白酶参与了无氧呼吸过程，使得微生物的胞外呼吸途径研究变得非常复杂。

希瓦氏菌是兼性厌氧菌，较容易培养，在厌氧条件可以利用金属氧化物［包括 Fe(Ⅲ)、Cr(Ⅵ)、U(Ⅵ)、Mn(Ⅲ) 及 Mn(Ⅳ) 矿物氧化物］、延胡索酸盐、硝酸盐、氧化三甲胺、硫酸二甲酯、亚硫酸盐、硫代硫酸盐以及单质硫等作为胞外电子受体（Myers and Nealson，1988），具有如此多样的胞外电子受体是其他任何胞外呼吸菌所不具备的。地杆菌被认为是环境中最主要的铁还原类群，可以利用有机物/有机污染物、乙酸盐、H_2 等作为电子供体，除了可以还原 Fe(Ⅲ) 外，还可还原重金属（Williams et al.，2011）。

1.2.2.2　微生物胞外呼吸电子传递过程

（1）胞内电子传递过程

随着希瓦氏菌属全基因组序列的测定和分析，对胞内电子传递过程的认识已经深入到分子水平。胞内电子传递过程中的第一步是脱氢酶从电子供体脱下电子，传递给醌类中间体。胞内电子传递过程中的第二步是电子从醌类中间体传递给 CymA（位于内膜多血红素细胞色素 c，是电子通过醌向周质传递的切入点）。胞内电子传递过程中的第三步是电子从 CymA 传递至周质细胞色素，目前已经发现的周质细胞色素主要是四血红素黄素细胞色素（Ifc3）和四血红素细胞色素（Cct）(Dobbin et al.，1999；Reyes-Ramirez et al.，2003)。胞内电子传递过程中的第四步是电子由周质细胞色素向外膜蛋白传递，目前发现镶嵌在周质和外膜上的色素蛋白（*MtrA*，位于周质，是可溶性的细胞色素 c，含有 10 个血红素）是这一过程的主要电子受体，*MtrA* 缺失将导致细胞与胞外电子受体之间的电子传递下降 90% 以上（Beliaev et al.，2001；Myers and Myers，2002；Pitts et al.，2003）。在还原可溶性电子受体时，与 *MtrA* 序列同源性较高的 *MtrD* 和 *DmsE* 可以部分代替 *MtrA*，但还原不溶性电子受体时 *MtrD* 和 *DmsE* 则不能代替 *MtrA*（Coursolle and Gralnick，2010）。胞内电子传递过程中的第五步是电子从 *MtrA* 向胞外传递，即外膜电子传递。尽管目前关于外膜电子传递的机制还不甚清楚，但有一个共同的认识是，无论是将电子直接传递至不同的电子受体或是传递至可溶性的电子穿梭体，外膜蛋白细胞色素 c(OM c-Cyt) 在这一过程中扮演着至关重要的角色（Lovley et al.，2011）。MtrB（非细胞色素，预测为跨膜蛋白）接受 *MtrA* 的电子，并传递给外膜的 OmcA 和 MtrC（位于外膜表面，均为脂蛋白，每个多肽包含 10 个血红素），后两者通常被认为是希瓦氏菌胞外电子传递的末端还原酶，它们经常以 2∶1 的复合体出现。体外实验表明，OmcA 和 MtrC 单独作用时也可以还原不溶性三价铁，但二者结合后的复合体的电子传递能力远大于单独作用时的电子传递能力（Shi et al.，2006）。OmcA 或 MtrC 的缺失都会导致产电量的大幅下降，但两者共同突变不会导致产电的完全消失，表明产电过程中存在多种的电子传递机制共存的现象或者 OmcA 或 MtrC 并不是唯一的末端还原酶（Coursolle and Gralnick，2010）。地杆菌属的胞内电子传递机制与希瓦氏菌属类似，但在具体细胞色素类型上有所不同，一般认为地杆菌属细胞质膜蛋白酶为 MacA，周质蛋白酶为 PpcA，外膜终端还原酶为 OmcB、OmcC、OmcE、OmcS、OmcT 与 OmcZ（Inoue et al.，2010）。此外，在地杆菌属细胞外膜上还发现一对了解不多但十分重要的多铜蛋白酶 OmpC 和 OmpB，它们对 Fe(Ⅲ) 的还原和电子转移到阳极都是非常重要的。

所有 OM c-Cyt 都呈现出类似的结构，由具有氧化还原活性的血红素分子和氨基酸组成的肽链缠绕形成，其中血红素的中心位置 Fe 轴向配位了双组氨酸残基接入肽链中（Clarke et al.，2011；Leys et al.，2002；Richardson et al.，2012；Shaik，2010）。这些具有氧化还原活性中心的蛋白位于细胞膜上，是电子从胞内到胞外的通道，这些蛋白中暴露于溶剂中的血红素分子在电子传递过程中可能会直接接触电子受体分子或界面，但其中的相互作用机制还有待研究。

Shewanella 属包含编码为 mtfDEF-omcA-mtfCAB 的基因族（Coursolle and Gralnick，2010；Shi et al.，2007）。MtrA 与 MtrB 组成了跨外膜的电子传递复合体，包括 β-折叠的孔蛋白（MtrB）以及嵌在其中的十血红素辅基细胞色素（MtrA）（Hartshorne et al.，2009；Ross et al.，2007）。MtrC 则是一个胞外十血红素辅基细胞色素，作为这个复合体的终端。MtrF、MtrD 与 MtrE 分别是 MtrC、MtrA 与 MtrB 的同源蛋白。操纵子 mfrDEF 在生物膜的生长中呈现出最高表达（McLean et al.，2008），但是会在 MtrCAB 和 MtrFDE 之间形成杂化复合体（Bücking et al.，2010；Coursolle and Gralnick，2010）。OmcA 蛋白是 MtrC 和 MtrF 的同源蛋白，能够与 MtrC 或 MtrF 相互作用，从 MtrCAB 或 MtrFDE 复合体接受电子（Shi et al.，2006），也可以在 ΔmtrC 或 ΔmtrF 突变株中替代这些蛋白（Coursolle and Gralnick，2010）。

目前已经对其中外膜电子传递通道中的一种十血红素辅基细胞色素（MtrF）进行了 X 射线晶体结构解析。根据这个结构模型，可以研究不同类型的胞内电子传递或解析可能的胞内电子传递发生机制。MtrF 晶体结构的解析第一次确定了 10 个血红素的空间排布构型，其中血红素以一种独特的交叉构型贯穿在四个结构域

(Domains Ⅰ、Ⅱ、Ⅲ、Ⅳ) 中 (Clarke et al.，2011)。这个结构可以为我们提供分子水平研究的可能性，用于分析胞外呼吸菌如何还原不溶性底物（如矿物）、可溶性底物（如黄素）以及与细胞表面不同氧化还原细胞色素终端之间形成的电子传递链 (Clarke et al.，2011)。

根据 MtrF 的晶体结构，Richardson 等（2012）提出了一种可能的外膜蛋白组成电子传递通道复合体的分子结构，由 MtrC、OmcA、MtrA 与 MtrB 构成，其中 MtrA 是基于两个五血红素辅基 NrfB 单体末端相连组成的 (Clarke et al.，2007)，MtrC 与 MtrA 嵌入孔蛋白的深度是未知的。目前实验方法还无法研究蛋白内沿着血红素组成的通道进行的电子传递，而高性能计算则可以从分子水平解析血红素分子之间电子传递的热力学和动力学性质 (Breuer et al.，2014)。采用这些蛋白组成独特的分子机器进行长距离的电子传输，转移电子的距离可以超过 100Å（1Å=10^{-10}m，下同），这对于生物纳米技术设备的设计具有明显的科学意义。

虽然目前对于胞内电子传递过程已经有很充分的探索，但是对于该过程涉及的界面过程和控制因素仍然很不清楚，而计算化学可以为实验结果提供充分而有力的证明与解释，能够为实验研究提供有指导价值的成果。在电子传递过程中可以对电子给体分子与电子受体分子以及固相界面组成的体系进行量子化学计算 (Vondrasek et al.，2005)，可以获得体系分子的电子结构与几何结构信息，从而分析分子轨道能级、原子的电子密度分布、电子传递反应过程的热力学和动力学性质 (Zeng et al.，2009)。自然体系中的电子传递过程一般发生在水溶液环境中，而分子动力学模拟可以构建较大的模型体系，以考虑真实环境体系中水溶液中近程水分子的影响，分析分子间作用力以及离子或分子在液体或固体中的扩散。通过有效的预测可以揭示其内在规律性。理论计算可以得到实验技术无法或者较难测定的物化性质。将计算模拟方法与精湛的实验技术有效结合，能够为研究提供更具优势的工具，加快研究成果的实际应用。

（2）胞外电子传递过程

微生物胞外呼吸菌将细胞代谢产生的电子从胞内传递到细胞外膜后，可以通过直接接触机制、纳米导线机制、应电运动机制和电子穿梭机制将电子转移给胞外电子受体（图1-1）。

实际环境介质中，各种微生物以群落形式存在，在它们所形成的代谢及呼吸网络中，上述各种电子传递机制将同时存在并相互协调。例如，非产电细菌可能会分泌某些电子穿梭体促进产电细菌的电子传递过程，而产电细菌的不完全代谢产物可能会为非产电细菌提供营养物质，细胞与细胞之间的纳米导线也可能会促使细胞之间发生电子传递。

图 1-1 中，①为好氧微生物的电子转移途径；②为硝酸盐呼吸菌和硫酸盐呼吸菌的电子转移途径；③~⑦为胞外呼吸菌的电子转移途径，其中③为胞外电子的直接接触传递方式，④为胞外电子的应电运动传递方式，⑤为胞外电子的纳米导线传递方式，⑥与⑦为胞外电子的电子穿梭体传递方式，⑥中的电子穿梭体在胞外环境介质中可移动，而⑦中的电子穿梭体在胞外环境介质中不可移动。

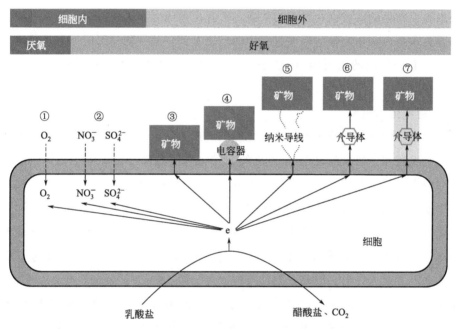

图 1-1　微生物细胞内代谢产生的电子的转移途径

1) 直接电子传递机制

直接电子传递机制是通过外膜上的活性蛋白将电子直接传递给电子受体，这类蛋白主要是末端还原酶和外膜表面黏性蛋白，目前研究发现希瓦氏菌属和地杆菌属多使用这种方式（Shi et al., 2006; Inoue et al., 2010）。微生物细胞外膜与电子受体的接触是直接电子传递的前提条件，也是其限速步骤，任何与接触蛋白有关的因子都会影响到直接电子传递的效率。直接电子传递机制与电子受体种类有关，例如，希瓦氏菌外膜终端还原蛋白 OmcA 或 MtrC 都可以还原胞外的核黄素类物质促进产电，但在 ΔmtrC 突变株可以表现出与野生型相当的核黄素还原能力，由此认为 OmcA 在核黄素还原过程中起到主要作用（Coursolle and Gralnick, 2010）。希瓦氏菌通过直接电子传递方式不仅可以将电子传递给可溶性的胞外电子受体，而且还可以传递给固相的胞外电子受体［例如Fe(Ⅲ)氧化物］，但与固相电子受体产生直接接触的外膜蛋白主要是MtrC（White et al.，2013）。同时，研究还发现，只要有连续的电子供应，微生物就完全能够在厌氧环境中通过直接接触固相铁矿物的方式维持生命（White et al.，2013）。理论上，在胞外直接电子传递的过程中会导致大量的胞外物质的产生，希瓦氏菌生物膜胞外物质主要以蛋白和多糖类等物质（Cao et al.，2011）为主，这些不导电物质的产生可能会阻碍生物膜细胞与电子受体的有效接触，从而抑制直接电子传递过程。此外，直接电子传递方式与细胞生物膜的形成也存在密切关系，希瓦氏菌的直接电子传递方式所必需的细胞外膜色素蛋白在生物膜的形成过程中会显著上调，其生物膜形成能力及其构成是影响直接电子传递的重要因素（Newton et al.，2009），也是一个容易被忽视的方面。

2）应电运动机制

某些微生物胞外呼吸菌可以将氧化底物所产生的电子储存在细胞表面，形成所谓的生物电容器，然后通过接触-传递的方式将电子转移给胞外电子受体，或者通过细胞鞭毛瞬间触及胞外电子受体的方式将电子释放，并迅速脱离电子受体表面，参与下次循环的电子传递。这种电子传递机制与电子穿梭机制有着明显的不同，无需电子穿梭体，是依靠微生物本身的应电运动方式传递电子。目前已经确定能够产生应电运动方式传递电子的微生物主要是希瓦氏菌，包括MR-1、SB2B与CN32，尤其是MR-1更为突出。研究发现，三种希瓦氏菌ΔmtrA、ΔmtrB与ΔcymA突变株都无法通过应电运动方式向胞外电子受体传递电子，这说明细胞外膜蛋白MtrA、MtrB与CymA在应电运动电子传递过程中是必不可少的部件。然而，并不是任何情况下都能够发生应电运动电子传递，合适的胞外电子受体才可以激发应电运动的发生，但这种应电运动又不同于微生物的趋药性和趋电性。研究表明，当MnO_2和微生物燃料电池的石墨电极作为胞外电子受体时，只能激发很少一部分的希瓦氏菌以应电运动方式进行传递电子，而当加入可溶性的胞外电子穿梭体（例如AQDS）后，以应电运动方式进行胞外电子传递的希瓦氏菌数目将大大增加，从而加速了MnO_2的还原和增加了微生物燃料电池的电流，这是由于在这种体系下希瓦氏菌首先是将电子以应电运动方式传递给电子穿梭体，然后再通过电子穿梭体传递给最终受体，而可溶性的电子穿梭体可能比固相的MnO_2和电极更能够激发应电运动的发生。不仅如此，可溶性的电子穿梭体也可以激发希瓦氏菌ΔmtrB突变株产生应电运动（Harris et al.，2009），但这与希瓦氏菌ΔmtrB突变株无法还原AQDS电子穿梭体（Lies et al.，2005）似乎是相矛盾的，关于其确切的原因目前还不甚清楚。

3) 纳米导线机制

纳米导线（nanowire）电子转移机制最初是在地杆菌 *G. sulfurreducens* 中发现的，它是指一定条件下微生物形成类似菌毛的导电附属肢体，这种导电附属肢体被称为纳米导线（Reguera et al.，2005），其作为电子导管可远距离向胞外电子受体传递电子，从而克服了细胞表面无法与电子受体直接接触的问题。通常情况下，纳米导线的直径只有 3~5nm，长度是直径的 1000 多倍，且非常耐用。除地杆菌之外，希瓦氏菌 MR-1、光合蓝绿菌中集胞藻（*Synechocysti*）和喜温发酵菌（*Pelotomaculum thermopropionicum*）在一定条件下也可以产生纳米导线（Gorby et al.，2006），表明纳米导线不只是异化金属还原菌的专属物，而可能是细菌有效获得电子的共同策略（见图 1-2）。希瓦氏菌纳米导线与地杆菌纳米导线在结构和组成上都存在差别，希瓦氏菌纳米导线呈电缆状，由多束更纤细的丝状物组成一股较粗的菌毛状结构，而地杆菌纳米导线则是呈单根状的表面附生结构，菌毛较细（Gorby et al.，2006）。然而，最近通过对希瓦氏菌 MR-1 活体细菌进行荧光测量、免疫标记以及定量基因表达分析，发现 MR-1 的纳米导线并非是之前一直认为的鞭毛结构，而是细胞外膜和周质的延伸，而且同样承载有 CymA、MtrA、MtrB、MtrC 和 OmcA 蛋白酶（Pirbadian et al.，2014）。

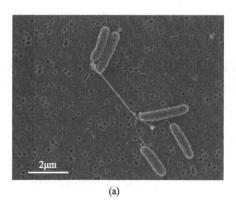 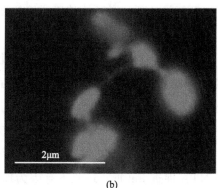

(a)　　　　　　　　　　　　　(b)

图 1-2　纳米导线

(Gorby et al.，2006)

纳米导线可以进入土壤和沉积物的纳米孔隙，不仅有利于细胞吸附于不溶性电子受体表面，而且可以传递电子进行还原作用（Reguera et al.，2005）。原子力扫描隧道显微镜证明了纳米导线具有较强的导电性能（Reguera et al.，2005），最近研究显示细菌纳米导线传输电子可达厘米级范围，是细菌自身大小的数千倍（Malvankar et al.，2011），说明相对于其他电子传递方式，纳米导线具有更优良的导电性。纳米导线主要是通过类似金属导电的形式或通过细胞色素间电子跃迁的形式进行传输电子（Malvankar et al.，2012）。前人研究发现细胞外膜蛋白OmcA或MtrC缺陷型的希瓦氏菌仍然可以产生纳米导线类似结构，但不具有导电性（Gorby et al.，2006），表明OmcA或MtrC是希瓦氏菌纳米导线传递电子所必须的，但关于其具体作用目前尚不清楚，有可能是参与的细菌纳米导线的构成。研究发现培养于不溶性铁氧化物的 *G.sulfurreducens* 可产生纳米导线，而可溶性的铁溶液中却没有此现象（Reguera et al.，2005），这显示微生物纳米导线的产生是可控制的。Reguera等（2005）对编码纳米导线蛋白亚组的基因GSU1496进行了检测实验，发现当缺乏该基因时，*G.sulfurreducens* 就不能产生纳米导线且不再还原不溶性电子受体，由此可见，通过转基因技术及环境诱导，使微生物长出纳米导线是可能的。此外，细菌纳米导线的导电性能还与环境条件存在密切关系，尤其是温度，这可能是由于温度过高会导致纳米导线产生无序的结构（Malvankar et al.，2011）。

微生物纳米导线还可以促进微生物燃料电池产电以及生物膜的形成，甚至可能形成细胞与细胞之间的链接及电子传递网络。细菌纳米导线的发现改变了微生物控制电子传输的传统理解，有益于纳米电子技术的应用发展，今后可从遗传上修改细菌纳米导线结构或合成不同功能的纳米导线，应用于纳米电子设备、微生物燃料电池、能源污染处理、微环境传感器等领域（Malvankar and Lovley，2012，2014）。

4）金属配位体增溶机制

金属配位体增溶机制对于微生物异化金属是非常重要的。Fe(Ⅲ)氧化物可与许多螯合剂形成铁配位体，该配位体的存在：一是可以增加反应体系中Fe(Ⅲ)的生物可利用性；二是可以提高Fe(Ⅲ)与异化铁还原菌直接接触的概率，从而提高铁还原的速率。Arnold等（1988）在研究希瓦氏菌SP200还原赤铁矿的实验中发现，加入等浓度的氮三乙酸（NTA），铁矿物溶解速率可提高20倍。在腐殖质土壤中NTA促进Fe(Ⅲ)还原不是因为增溶铁，而是因为NTA的加入增加了从土壤腐殖质向水溶液中的溶解，进而通过溶解性的腐殖质作为电子穿梭体来加速铁(Ⅲ)还原。除NTA外，乙二胺四乙酸（EDTA）、乙醇二氨基乙酸（EDG）、六偏磷酸钠（Calgon）、甲基亚氨基二乙酸（MIDA）与多磷酸盐等螯合物也能够促进Fe(Ⅲ)氧化物的溶解及还原（Lovley et al.，1996）。同时，自然界中也存在多种铁螯合物，如麦芽糖醇与邻苯二酚等（Dobbin et al.，1999）。另外，有些微生物能分泌螯合物，如高铁载体（siderophores），它是微生物在缺铁的情况下分泌到细胞外的低分子量有机化合物，可与Fe(Ⅲ)进行配位而增溶铁。尽管上述这些螯合物改变了环境中铁的存在状态，促进了铁元素及其他元素的循环，但它们在环境中也可能受到其他物质或微生物的影响，使其作用力损耗，限制了对不溶性Fe(Ⅲ)的还原转化。

5）电子穿梭机制

上述各种电子传递方式只能在细胞与电子受体直接接触或者纳米级的距离内进行，而事实上，许多不溶性的电子受体的还原可在一定距离外进行，这暗示着微生物可以通过电子穿梭体物质还原不溶性电子受体。电子穿梭机制是指微生物利用自然环境中广泛存在的腐殖质、植物根系分泌物、或细胞自身合成的电子穿梭体（Lovley et al., 1996；Marsili et al., 2008；Okamoto et al., 2013），接受胞内传递出的电子，并将其运出细胞，传递给胞外电子受体后，以氧化态返回细胞再次接受电子，如此往返穿梭于细胞与电子受体之间的电子传递（图1-3）。

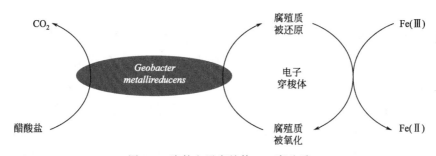

图1-3 胞外电子穿梭体——腐殖质

在环境中，细胞外的电子传递中介体被认为对可利用不溶性物质作为电子供体或受体的微生物是非常重要的（Watanabe et al., 2009）。一个典型实例是异化铁还原菌，尽管十几年前认为异化铁还原菌必须直接接触三价铁氧化物才能将它们作为电子受体，但实验证据已显示，无论是自然存在的、还是自身分泌合成的电子穿梭体，在细菌细胞和三价铁氧化物之间的长距离扩散上均可以通过电子转移而有效解决上述的局限性（Hernandez and Newman, 2001；Nevin and Lovley, 2002），甚至很低的电子穿梭体都会对已知环境中的末端氧化剂的电子转移产生显著的影响（Marsili et al., 2008）。此外，最近基于动力学的研究已经发现，诸如地杆菌 *Geobacter sulfurreducens* 之类的异化铁还原菌转运电子至电子穿梭体（如环境浓度的腐殖质）的速率远高于转运至固相氢氧化铁的速率（至少是27倍）(Jiang and Kappler, 2008)，由此表明电子穿梭体对于环境中异化铁还原菌的能量代谢具有重要意义。此外，基于电子穿梭体的电子转移过程在有机污染物生物修复、污水处理等环境治理方面也起到积极作用，例如胡敏酸作为常见的电子穿梭体，可以促进厌氧污泥高效去除苯酚和四氯化碳污染物（Martinez et al.，2013），核黄素通过促进电子传递可以加快环杂硝胺的降解。

（3）细胞间电子传递过程

微生物本质上来说并不会作为单纯的培养物存在，而是与邻近的细菌、古细菌及真核生物等共同参与各种相互作用和产生营养相互依赖性存在于生态系统中。微生物之间的互养共栖就是一种典型的营养相互依赖性的生存方式，其中两种或者多种微生物之间会组合新陈代谢的能力，以降解那些单个微生物所无法独立降解的特殊物质。两个微生物合作伙伴之间通过共享电子的互养共栖对于各种产甲烷环境的生态功能运转是非常重要的，例如在湿地、水中沉积物、储油池以及将有机垃圾转换成甲烷的厌氧分解池。微生物细胞间的电子传递被认为在微生物聚集体中起到了非常重要的作用，这些微生物聚集体可以通过还原硫酸盐的方式来厌氧氧化甲烷。

细胞间电子传递最著名的策略就是 H_2 细胞间电子传递体，其中供电子的微生物会将质子还原产生 H_2，而产甲烷菌则通过氧化 H_2 将二氧化碳还原成甲烷。在某些情况下，甲酸盐会取代 H_2 作为细胞间电子传递体。在厌氧土壤、沉积物以及某些厌氧分解池中存在有丰富的依靠消耗 H_2/甲酸盐生存的产甲烷菌，这表明 H_2/甲酸盐细胞间电子传递体对于调控这些环境的甲烷生成起到了非常重要的作用。目前，在 H_2 供体微生物与 H_2 消耗产甲烷菌的混合培养研究中已经对 H_2 细胞间电子传递体做了详细描述，而且对 H_2 产生及 H_2 消耗的生理学与生物化学过程也有较深刻的理解。

另外一种代替 H_2 细胞间电子传递体的机制就是细胞间直接电子传递。甲烷丝状菌（*Methanosaeta haurindacaea*）是一种地球上最重要的产甲烷的微生物，其在混合培养中会直接接受来自地杆菌（*Geobacter metallireducens*）的电子，将二氧化碳还原成甲烷。有多个证据表明，甲烷丝状菌在处理啤酒厂垃圾的厌氧分解池中也会参与细胞间的直接电子传递；在分解池的颗粒中也富含地杆菌种，其拥有类似于金属的导电性能，这与地杆菌菌毛的导电性非常相似（Rotarua et al.，2014）。由此可见，在甲烷丝状菌和地杆菌混合培养中，地杆菌产生的菌毛在细胞间直接电子传递机制中起到十分关键的作用，其主要是通过生物膜上的氧化还原辅因子所形成的碳-碳连接结构进行多级电子跃迁，从而实现电子传递（Nagarajan et al.，2013）。

乙酸产甲烷菌中的甲烷八叠球菌（*Methanosarcina*）也具有细胞间直接电子传递机制，它在产甲烷土壤、沉积物、煤矿、垃圾填埋场及厌氧分解池中也都有广泛的分布（Hu et al.，2013）。研究表明，甲烷八叠球菌可以接受来自细胞外非生物表面体的电子（Kato et al.，2012），例如，甲烷八叠球菌（*M. barkeri*）会吸附到颗粒活性炭上接纳电子，这种颗粒活性炭可以作为 *M. barkeri* 与 *G. metallireducens* 之间电子传递的中介体（Liu et al.，2012）。缺乏菌毛或与菌毛有关的细胞色素 OmcS 的地杆菌 *G. sulfurreducens* 突变体是无法在没有颗粒活性炭的厌氧环境中进行细胞间直接电子传递，而在富含颗粒活性炭的情况下却可以很好地进行细胞间直接电子传递（Liu et al.，2012），表明颗粒活性炭在细胞间形成了电子流通道。

1.2.2.3 微生物胞外呼吸与铁（Ⅲ）还原

Fe(Ⅲ) 可以被非生物还原，也可以被异化铁还原微生物还原。Fe(Ⅲ) 的还原会耦合 Fe、C 与 S 等元素的循环，这几乎在自然界的所有环境中都会发生。

很多微生物都可以通过各类电子供体还原 Fe(Ⅲ)，如醋酸盐、乳酸盐和 H_2。其中最常见 Fe(Ⅲ) 还原菌包括 *Geobacter* spp.、*Shewanella* spp.、*Albidoferax ferrireducens*、*Geothrix fermentans* 以及各类超嗜热古菌（Lovley and Phillips，1988；Lovley et al.，2011；Myers and Nealson，1990；Obuekwe et al.，1981；Ramana et al.，2009；Coates et al.，1999）。*Geobacter* spp. 是第一个被证明能够通过短链脂肪酸、单环芳香化合物（如甲苯或者苯）或者氢作为电子供体的 Fe(Ⅲ) 还原菌（Lovley et al.，1987；Lovley et al.，2011；Zhang et al.，2013），将 Fe(Ⅲ) 矿物还原成含有混合价态的磁性矿物 Fe_3O_4。

与发酵细菌一起，*Geobacter* spp. 可以将有机碳完全矿化成 CO_2，这在还原性环境中非常普遍（Lovley et al.，2011）。*Geobacter* spp. 对由 Fe(Ⅲ) 矿物还原产生的 Fe(Ⅱ) 会表现出趋药性行为，并通过这种方式找到 Fe(Ⅲ) 矿物源，然后利用菌毛吸附在矿物表面上（Childers et al.，2002；Methé et al.，2003）。除了还原 Fe(Ⅲ) 之外，*Geobacter sulfurreducens* 还会在醋酸盐存在的情况下，将 NO_3^- 还原成 NH_3 的过程与 Fe(Ⅱ) 氧化过程耦合在一起（Coby et al.，2011；Methé et al.，2003）。

另外一种研究比较多的 Fe(Ⅲ) 还原菌是希瓦氏菌，尤其是 *S. oneidensis* MR-1，它是于 20 世纪 90 年代被发现的（Myers and Nealson，1990），可以用氢、甲酸盐或者乳酸盐作为电子受体来还原 Fe(Ⅲ)。

Shewanella spp. 与 *Geobacter* spp. 的基因组序列信息有助于确定与 Fe(Ⅲ) 还原途径有关的基因（Heidelberg et al.，2002；Methé et al.，2003）。来自细胞内

分解代谢的电子会被转移到细胞局部表面的 c 型细胞色素上，这些细胞色素可以催化细胞外电子转移来还原 Fe(Ⅲ) 与 Mn(Ⅳ) 氧化物（Shi et al.，2009）。在 *S. oneidensis* MR-1 中，与外膜有关的 c 细胞色素 MtrC、MtrF 及 OmcA 被认为对绑定和还原 Fe 矿物都起作用（Donald et al.，2008；Reardon et al.，2010；Shi et al.，2009）。外膜细胞色素还会通过外膜乳蛋白-细胞色素复合物与细胞内醌池的呼吸电子之间建立联系，比如说 MtrA、MtrB 与 CymA（White et al.，2013）。MtrA 与 MtrB 的同系物在很多变形菌门的微生物中都可以得到系统发育性地继承保存，包括 *Shewanella*、*Geobacter* 与 *Rhodopseudomonas* 属（Hartshorne et al.，2009；Jiao and Newman，2007；Shi et al.，2012）。很多地杆菌属都会分泌细胞外细胞色素，例如 *G. sulfurreducens* 分泌的 OmcS 能够与导电性的菌毛纳米导线结合，从而可以调节纳米导线上电流的传导，或者作为还原 Fe(Ⅲ) 矿物的接触点（Qian et al.，2010；Leang et al.，2010；Malvankar et al.，2011）。除了细胞色素之外，*G. sulfurreducens* 还需要外膜乳蛋白 OmpJ 用于 Fe(Ⅲ) 的还原（Afkar et al.，2005）。尽管 *S. oneidensis* 与 *G. sulfurreducens* 在电子传输途径上含有相类似的元件，但它们的功能是完全不同的（Shi et al.，2007）。此外，不同的 *Geobacter* 菌种似乎具有不同的细胞色素附件（Butler et al.，2010），这说明 Fe(Ⅲ) 还原所必需的电子转移可以通过各种不同的生物化学途径来实现。

由于 Fe(Ⅲ) 氢氧化物的溶解性很差，并且细胞色素之间电子能够"跃迁"的最大距离只有 2.0 nm（Gray and Winkler，2009），因此，理解 Fe(Ⅲ) 矿物的微生物还原一般都会涉及电子从微生物细胞向外转移到 Fe(Ⅲ) 矿物的策略，而不仅仅是细胞-矿物直接接触的策略（Hernandez and Newman，2001）。目前，有几种机制可以被用来解释电子从细胞转移到胞外固体表面的过程，这几种机制可以在微米、毫米甚至厘米级的距离上进行电子传输（Hernandez and Newman，2001；Nielsen et al.，2010；Shi et al.，2012）。在低 Fe(Ⅲ) 环境下，希瓦氏菌 *S. oneidensis* 会分泌具有氧化还原活性的电子穿梭体（例如黄素），搭建细胞与 Fe(Ⅲ) 矿物之间的电子传输（Marsili et al.，2008；von Canstein et al.，2008），或者利用 Fe(Ⅲ) 螯合剂，进而促进 Fe(Ⅲ) 作为电子受体（Nevin and Lovley，2002）。在 *Shewanella* 与 *Geobacter* 的胞外电子转移中还会牵涉具有氧化还原活性的菌毛纳米导线的产生（Gorby et al.，2006；Reguera et al.，2005）。此外，*Shewanella* 与 *Geobacter* 物种也会通过外源的电子穿梭体的策略将电子转移到远离细胞所在位置的 Fe(Ⅲ) 矿物上（Lies et al.，2005；Rosso et al.，2003），这种电子穿梭体也是一种具有氧化还原活性的分子成分，例如硫化合物和腐殖质等天然有机物（Lohmayer et al.，2014）。

溶解性与固相腐殖质都可以刺激Fe(Ⅲ)矿物的微生物还原（Lovley et al.，1996；Roden et al.，2010）。该过程的第一步会牵涉微生物将电子供给腐殖质（包括胡敏酸和富里酸），第二步是还原的腐殖质通过非生物电子供给的方式将电子传递给Fe(Ⅲ)矿物。腐殖质的这种能力不只限于异化金属还原菌，湖泊与海洋沉积物以及无污染与被污染的湿地沉积物中的发酵细菌、产甲烷菌和硫酸盐还原菌也都具有将电子传递给腐殖质的能力（Benz et al.，1998；Cervantes et al.，2002）。因此，在无法采用微生物酶来直接还原Fe(Ⅲ)的情况下，基于腐殖质作为电子穿梭体的Fe(Ⅲ)非生物还原是一种促进生物地球化学铁循环的重要途径（Piepenbrock et al.，2014）。

在很多情况下，硫化合物可以从微生物膜传递出的电子，然后来实现Fe(Ⅲ)矿物的还原（Yao and Millero，1996）。在中性pH下，微生物还原后产生的H_2S可以通过非生物的方式来还原Fe(Ⅲ)氧化物，其还原速率取决于矿物质表面积以及pH值，而且还原速率还会与H_2S及Fe(Ⅲ)氧化物浓度之间表现出一阶动力学关系（Yao and Millero，1996）。在海洋环境中，高浓度的硫酸盐和元素硫的微生物还原会产生大量的H_2S，在这种情况下，H_2S与Fe之间的反应是非常重要的，因为该反应过程会将挥发性的H_2S形成沉淀形式的硫单质。因此，在某种程度上来说，铁矿物可以通过防止H_2S从沉积物中挥发到上覆水中的方式来控制H_2S的分布（Canfield，1989；Thamdrup et al.，1994）。

1.2.2.4 微生物胞外呼吸在环境污染修复中的应用

人类活动产生的有机和无机污染物造成了大量环境问题。多数情况下，污染物的行为是由复杂的生物地球化学过程控制，而微生物群落则可以调节这些生物地球化学过程。胞外呼吸菌，尤其是异化金属还原菌，其胞外电子传递过程是最近新发现的新型微生物厌氧能量代谢方式，它作为一种独特的氧化还原过程，在污染物原位修复、污水处理以及清洁生物能源提取方面已经逐渐呈现出不可替代的优越性和重要的应用前景，因此近几年来受到科学界的广泛关注。

(1) 有机污染物的氧化与还原降解

人类活动产生的有机物污染物通常含有芳香族化合物，而苯环的热稳定性就使得这些污染物难以转化，长期存留于环境中（Carmona and Díaz，2005）。大量有机物电子供体的异化作用会把氧气作为电子受体而迅速消耗掉，从而形成厌氧环境

(Carmona et al., 2009)，这为 Fe(Ⅲ) 氧化物作为电子受体参与有机物的氧化提供了主要场所。经过对环境的现场检测已证明，Fe(Ⅲ) 还原时会促进有机物降解 (Lyngkilde and Christensen, 1992; Nielsen et al., 1995; Christensen et al., 2001)。最先被发现的能够将 Fe(Ⅲ) 还原同时又可氧化芳香族化合物的单一菌属是异化金属地杆菌属 *G. metallireducens* (Lovley et al., 1989)，这一菌属是从烃类污染土壤中分离出来的，它可以利用 Fe(Ⅲ) 当作电子受体，从苯甲酸酯、甲苯、苯酚和对甲苯酚的氧化过程中获得能量 (Lovley and Lonergan, 1990)。它还能够将芳香族化合物矿化，产生 CO_2，并产生极少的中间产物。对于硫还原地杆菌的基因组的研究，读者可参考 Wischgoll 等 (2005) 和 Carmona 等 (2009) 的研究。迄今为止还没有研究显示地杆菌属某单一菌株能够将 Fe(Ⅲ) 还原并氧化芳香族化合物 (Butler et al., 2007)，但仍有发现 *Geobacter hydrogenophilus* 和 *Geobacter grbiciae* 两种地杆菌同时存在时，两者均可以将苯甲酸盐氧化，后者甚至还可以氧化甲苯 (Coates et al., 2001)。对剔除了地杆菌属基因后的细菌进行纯培养时，没有发现其他菌属具有将 Fe(Ⅲ) 还原并与芳烃氧化耦合的能力。最新的研究又分离出了两种带有这种新陈代谢特性的细菌，分别是 *Geobacter toluenoxydans* 和 *Desulfitobacterium aromaticivoran*，其中后者是属于革兰氏阳性梭状芽孢杆菌 (Kunapuli et al., 2010)。

虽然目前发现的地杆菌属能够将 Fe(Ⅲ) 还原并与芳香族化合物氧化耦合的并不多，但在沉积环境中确实有丰富的地杆菌，而且其中大部分都具有 Fe(Ⅲ) 还原能力，说明这一研究还是有其环境意义的 (Coates et al., 1996)。利用 16S rDNA 序列分析等微生物生态学技术，发现这些地杆菌对芳烃污染的土壤修复意义重大 (Rooney-Varga et al., 1999; Roling et al., 2001; Staats et al., 2011; Snoeyenbos-West et al., 2000)，说明利用这些细菌进行生物强化来促进生物降解，制定一个较好的生物修复方案是可行的。然而，目前存在的主要问题是，微生物如何接触到固态的铁氧化物 (Lovley et al., 1994; Lovley et al., 1996)。Fe(Ⅲ) 还原与芳烃氧化耦合的过程会产生活性较强的 Fe(Ⅱ)，例如最新的研究表明，地杆菌属 *G. metallireducens* 氧化 BTEX（苯、甲苯、乙苯和二甲苯）时所产生的 Fe(Ⅱ)，可以进一步介导酶污染物和硝基化合物的还原 (Tobler et al., 2007)，这表明，微生物还原 Fe(Ⅲ) 与芳烃氧化耦合过程可以将那些微生物无法直接利用的共存污染物进行降解 (Tobler et al., 2007)。

如上所述，微生物还原 Fe(Ⅲ) 产生的 Fe(Ⅱ) 可以通过非生物电子交换反应还原转换污染物。事实上，在工业生产以及污染物的化学修复中，Fe(Ⅱ) 是很常用的还原剂（Charlet et al.，1998）。生物 Fe(Ⅱ) 通常是由含铁的生物矿物产生，这些 Fe(Ⅱ) 或是存在于矿物结构中，或是吸附在矿物表面（Cutting et al.，2009）。已有实验证明，Fe(Ⅱ) 矿物参与了六氯乙烷的脱卤和硝基氯苯的硝基还原（Elsner et al.，2004），还发现随着铁(Ⅱ)矿物成分不同，矿物表面的反应速率也不同，其中，铁硫化合物较其他矿物反应速率会比较大。在地下环境系统中，磁铁矿[也是由微生物还原 Fe(Ⅲ) 产生的]是一种很常见的物质，它能与污染物发生反应，这已经引起了许多关注（Gorski and Scherer，2009；Gorski et al.，2010）。Gorski 等（2010）还发现，当磁铁矿中 Fe(Ⅱ) 的含量增加，硝基苯反应速率和铁矿合成的反应速率都会加快。McCormick 和 Adriaens（2004）将磁铁矿应用到了 CCl_4 的转化中，发现反应过程中约有 47% 的 CCl_4 经过脱卤反应转化成了 CO_2 和 CH_4，但是关于 CH_4 的产生机理尚不明确。生物磁铁矿与微量的菱铁矿或某种含 Fe(Ⅱ) 物质相结合可以促进地下水中环三亚甲基三硝胺（RDX）的转化，虽然生成量只占反应物的 30%，但却不断有亚硝基的生成，并最终导致 1,3,5-三甲基己羟基-1,3,5-三嗪的积累（Williams et al.，2005）。Borch 等（2005）把纤维菌 *Cellulomonas* sp. 放在含 TNT 的基质中培养，并加入水铁矿与电子穿梭体 AQDS，实验结果发现，基质中迅速产生了 TNT 还原产物，而且这些产物比亲体分子更不易移动。

（2）重金属的还原转化

1）Cr(Ⅵ) 的还原

Cr 的氧化还原活性很强，环境中主要以 Cr(Ⅵ) 和 Cr(Ⅲ) 两种价态存在。在酸性和中性条件下，主要以 Cr(Ⅲ) 的形态存在，而在碱性和氧化性环境中主要以 Cr(Ⅵ) 的形态存在（Kimbrough et al.，1999）。环境中 Cr(Ⅲ) 主要以铬氧化物的形式存在，通常与铁结合形成铬铁矿或吸附在矿物表面，只有在酸性很强的条件下才能够被溶解（Fendorf，1995）。Cr(Ⅵ) 的存在形式有很多，主要有 H_2CrO_4、$HCrO_4^-$、CrO_4^{2-} 和 $Cr_2O_7^{2-}$ 等，都是易溶物且不易吸附在矿物表面（Kimbrough et al.，1999）。

Cr(Ⅵ) 是有毒的（Chen and Hao，1998），而 Cr(Ⅲ) 没有毒性，还是糖脂代谢必需的营养物质（Wang，2000）。传统的处理 Cr(Ⅵ) 污染物的方法就是化学还原为 Cr(Ⅲ) 氧化物，然后沉淀去除。由于 Cr(Ⅵ)/Cr(Ⅲ) 氧化还原电位较高，溶解氧对 Cr(Ⅲ) 的再氧化在动力学上是非常缓慢的（Rai et al.，1989）。尽管水体和土壤中存在锰氧化物可以再氧化 Cr(Ⅲ)，但这是一个溶解控制过程，且进行得缓慢（Rai et al.，1989）。

虽然微生物可以通过胞外呼吸途径产生 Fe(Ⅱ) 或硫化物来间接促进 Cr(Ⅵ) 的还原,但研究发现微生物也可以通过直接电子传递方式来还原 Cr(Ⅵ)。假单胞菌 *Pseudomonas dechromaticen* 和 *Pseudomonas chromatophila* 是最早被发现具有独自进行酶催化还原 Cr(Ⅵ) 的能力 (Romanenko and Koren'kov, 1977; Lebedeva and Lialikova, 1979),随着不断地研究,发现了更多的有此特性的细菌 (Cervantes et al., 2007),大多为兼性厌氧菌。早期的研究普遍认为,细菌还原 Cr(Ⅵ) 的过程没有能量的存储 (Ishibashi et al., 1990),但随后就有研究指出,细菌在还原 Cr(Ⅵ) 的过程中也在不断地生长 (Tebo and Obraztsova, 1998; Francis et al., 2000)。此外,一些从土壤和水体中分离出的真菌也具有还原 Cr(Ⅵ) 的能力,例如 *Aspergillus* sp. N_2、*Penicillium* sp. N_3、*Hypocrea tawa* 与 *Paecilomyces lilacinus* (Barrera-Díaz et al., 2012)。

Cr(Ⅵ) 的微生物还原机制因酶在胞内位置的不同而异 (Cervantes et al., 2007)。细胞膜上的 Cr(Ⅵ) 还原与呼吸链及相关的细胞色素有关,然而细胞质中的 Cr(Ⅵ) 还原与黄素还原酶有关 (Magnuson et al., 2010)。在厌氧环境中,假单胞菌 *Pseudomonas ambigua* G-1 和 *Pseudomonas putida* 以 NAD(P)H 为电子供体,通过溶解 Cr(Ⅵ) 还原酶将 Cr(Ⅵ) 还原。正是基于这些实验研究,人们发现了越来越多溶解 Cr(Ⅵ) 还原酶也能够还原 Cr(Ⅵ),如 ChrR 和 NfsA (Ackerley et al., 2004; Barak et al., 2006)。不同的酶其还原机制不同,NfsA 在还原 Cr(Ⅵ) 时有两个电子转移 (Ackerley et al., 2004),产生的中间产物 Cr(Ⅴ) 会加快反应进行,而且反应生成的活性氧物质 (ROS) 与 Cr(Ⅵ) 的毒性有关 (Barak et al., 2006)。经过 *Pseudomonas aeruginosa* 还原 Cr(Ⅵ) 的蛋白质组学研究,发现 ROS 解毒蛋白的过表达产物确实能降低 Cr(Ⅵ) 的毒性 (Kilic et al., 2009)。大肠杆菌的可溶性还原酶 ChrR 在氧化还原反应过程中有 4 个电子转换,其中,3 个电子用于将 Cr(Ⅵ) 还原为 Cr(Ⅲ),还有 1 个电子传递到了分子氧,由此限制了 ROS 的产生 (Ackerley et al., 2004)。

一些专性厌氧菌还原 Cr(Ⅵ) 时会把 Cr(Ⅵ) 当作电子受体。硫酸盐还原细菌是受到较多关注的微生物,因为它既能氧化乳酸又能还原 Cr(Ⅵ),这与其还原硫酸盐和铬酸盐机理相似 (Lloyd et al., 2001)。硫酸盐还原细菌 *Desulfovibrio* spp. 可以把氢气作为电子供体,利用 c-细胞色素与氢化酶将 Cr(Ⅵ) 还原 (Lovley and Phillips, 1994)。在电子转移过程中,希瓦氏菌 *S. oneidensis* 可以利用 MtrC 和 OmcA 两种细胞色素作为终端还原酶将胞外 Cr(Ⅵ) 还原 (Belchik et al., 2011),而当去除相应的 *mtrC* 和 *omcA* 两组基因后,细菌则不再还原 Cr(Ⅵ),同时胞内 Cr(Ⅲ) 积累增加，胞外 Cr(Ⅲ) 减少（Belchik et al.，2011）。

将微生物对 Cr(Ⅵ) 的还原最终应用于微生物修复当中才是研究的目的。异位修复和污水处理系统的生物反应器均已实现了 Cr(Ⅵ) 的还原，这是由浮游细菌（Tripathi，2002）或是固定床生物膜上的细菌完成的（Konovalova et al.，2003）。Tripathi（2002）利用模式微生物 *P. aeruginosa* A2Chr 研究不同的生物反应器对 Cr(Ⅵ) 的还原，实验得知，与浮游细菌相比，固定床生物膜上的细菌不易受到 Cr(Ⅵ) 毒性的影响，因此可以应用于高浓度的 Cr(Ⅵ) 污染环境中。在固定床生物膜上接种 *Bacillus* spp.（Chirwa and Wang，1997）、硫酸盐还原菌（Smith，2001）或是混合接种（Nancharaiah et al.，2010）等研究都证实了以上结论，且在好氧和厌氧环境下都适用，但是，在缺氧状态下细菌对 Cr(Ⅵ) 的去除能力更强。

Cr(Ⅵ) 污染的原位处理也是一项重要的技术，土壤或沉积物中都有土著 Cr(Ⅵ) 还原菌（Bader et al.，1999）。通过各种不同的手段刺激细菌的代谢，例如，向电镀污染土壤中添加大豆胰蛋白酶、葡萄糖和矿物盐，发现细菌不仅还原了 Cr(Ⅵ)，而且还刺激了二氧化碳转化（Turick et al.，1998）。由于原位试验的环境条件不易控制，因此微生物代谢活动经常不会处于最佳状态，就大多数地下含水层来说，通常温度较低且电子供体有限。极端的酸碱度也是影响 Cr(Ⅵ) 污染原位处理的重要因素，因为 Cr(Ⅵ) 主要存在于碱性环境中。最近，很多实验研究从碱性环境中分离出了适应高 pH 值（通常 pH 9～10）的 Cr(Ⅵ) 还原细菌，如 *Alkaliphilus metalliredigens*（QYMF）(Roh et al.，2007)，*Burkholderia cepacia* MCMB-821（Wani et al.，2007）和 *Halomonas* sp.（Vanengelen et al.，2008）。对从不同的受 Cr(Ⅵ) 污染的样品中分离出的 Cr(Ⅵ) 还原细菌进行混合培养（Jeyasingh and Philip，2005；Jeyasingh et al.，2010），发现混合培养的 Cr(Ⅵ) 还原细菌能够将 Cr(Ⅵ) 完全还原。

2）汞的还原

汞的毒性取决于其价态，Hg(Ⅱ) 毒性最强，Hg(0) 毒性相对较小（Clarkson，1997）。受到最大关注的是甲基汞，这并不是因为细胞内的甲基汞进行脱甲基反应能生成有毒的 Hg(Ⅱ)，而是因为甲基汞更易移动（Morel et al.，1998）。

尽管自然界中汞的存在形态及其水平与许多化学过程有关（Jonsson et al.，2014；Zheng et al.，2013），但很多情况下降低或消除汞的毒性以及汞形态的转化都是通过微生物进行的（Hu et al.，2013；Schaefer and Morel，2009）。汞的剧毒性以及自然汞事件的频频发生使得许多细菌形成了一套能够解毒 Hg(Ⅱ) 的机制，包括细胞对 Hg(Ⅱ) 的吸收，胞内 Hg(Ⅱ) 被还原为毒性较小的 Hg(0)，最后基于高蒸气压和 Hg(0) 低溶解度被排出胞外（Barkay et al.，2005）。Hg(Ⅱ) 是通过主动运输进入细胞，这一过程需要一系列特定的蛋白参与，包括周质蛋白 MerP 和细胞质膜蛋白 MerT、MerC、MerF 和 MerE（Barkay et al.，2003）。一旦 Hg(Ⅱ) 进入到细胞内，就可以通过氧化还原中介体（谷胱甘肽、半胱氨酸）或直接与蛋白酶 MerA 结合（Barkay et al.，2003）。蛋白酶 MerA 是一种醌氧化还原酶，可将 Hg(Ⅱ) 还原为 Hg(0)，然后经被动扩散把 Hg(0) 运到细胞外（Barkay et al.，2003）。在厌氧环境中，被胡敏酸还原菌还原后的胡敏酸可以明显促进 Hg(Ⅱ) 的还原（Gu et al.，2011），说明 Hg(Ⅱ) 的还原会受到胞外呼吸过程的影响。

微生物对汞的解毒作用已经在污水处理试验（Wagner-Döbler，2003）和水、土壤及沉积环境的原位修复中得到了应用（Saouter et al.，1995；Nakamura et al.，1999）。在处理氯碱电解废水中，利用固定床生物反应器将污水中的 Hg(Ⅱ) 还原为 Hg(0)，然后反应器上密布的惰性载体（如浮石）和炭层过滤器会把流经过的 Hg(0) 吸附并固定（Wagner-Döbler，2003）。这种生物反应器上接种有耐汞微生物，主要是假单胞菌，这些细菌组成了一个厚的细胞生物膜和胞外多糖组织（Wagner-Döbler et al.，2000）。

汞污染土壤的原位生物修复也是一个重要的技术。Saouter 等（1995）从 Hg(Ⅱ) 污染的湖泊中分离出耐汞微生物 *Aeromonas hydrophila* KT20 并进行接种培养，结果发现 Hg(0) 生成量有所增加。解毒蛋白 *mer* 的过度表达也有可能提高微生物对 Hg(Ⅱ) 的去除效率（Brim et al.，2000），但不幸的是转基因微生物不能轻易的应用到环境中去（Lloyd et al.，2003）。Brim 等（2000）还对防辐射细菌 *Deinococcus radiodurans* 展开了基因工程研究，并将其应用于修复放射性核素和汞污染的环境，这一菌株中带有克隆的耐汞基因 merA，即使在强辐射环境中也能够将 Hg(Ⅱ) 还原为 Hg(0)。此外，某些金属还原细菌以及各种硫酸盐还原菌（King et al.，2000）也能代谢产生甲基汞，而且这还是缺氧沉积物中甲基汞的主要来源。

3) 铀的还原

U 在环境中主要以 U(Ⅳ) 和 U(Ⅵ) 氧化物的形式存在。U(Ⅳ) 氧化物可溶性差，而 U(Ⅵ) 碳酸盐复合物的可溶性很强，如 $UO_2(CO_3)_2^{2-}$ 和 $UO_2(CO_3)_3^{4-}$（Clark et al.，1995）。将铀还原可实现铀的固定（Williams et al.，2013），这是一种重要的污染修复机制，能够防止 U(Ⅵ) 迁移到地下水系统。

普遍认为在厌氧环境中铀的去除是一种间接还原，而且会牵涉许多微生物胞外呼吸的过程（Newsome et al.，2014）。例如，Lovley 等（1991）认为，微生物会产生 Fe(Ⅱ) 或硫化物等还原剂将铀还原。他们还用醋酸盐和 H_2 作为电子供体，研究异化金属还原菌 G. metallireducens 和 S. oneidensis 对铀的还原，发现两种菌直接催化还原 U(Ⅵ) 的同时自身也在进行代谢生长。紧接着，又有学者通过培养硫酸盐还原菌 D. desulfuricans 和 D. vulgaris，也得出了与上述相类似的结论（Lovley and Phillips，1992；Lovley et al.，1993）。尽管在实验中硫酸盐还原菌无法储存生长所需的能量，但 U(Ⅵ) 确实被迅速地还原为 U(Ⅳ)。最近研究表明，在电子供体碳源供应充足的情况下，U(Ⅵ) 的微生物还原与微生物量存在密切关系（Barlett et al.，2012a），地杆菌与硫酸盐还原菌对还原 U(Ⅵ) 不存在竞争关系（Barlett et al.，2012b），表明两种菌对还原 U(Ⅵ) 存在两种相互独立的机制。但无论如何，U(Ⅵ) 的微生物还原速率比非生物还原要快，这暗示着微生物还原 U(Ⅵ) 可以被应用到环境污染修复中去（Holmes et al.，2015）。研究发现，微生物 D. vulgaris 在还原 U(Ⅵ) 的过程中，c3 细胞色素起到了关键作用，这与 Cr(Ⅵ) 的微生物还原机理相似，而当用阳离子交换柱把电子传递蛋白从细胞的可溶性部分移除后，发现微生物就失去了还原能力（Lovley et al.，1993）。地杆菌 G. sulfurreducens 还原 U(Ⅵ) 的机理相对较复杂，因为在其细胞中有许多细胞色素，细胞周质（Lloyd et al.，2002）和细胞外膜（Shelobolina et al.，2007）中都有细胞色素，而且，最新研究还发现，胞外纳米导线也参与了 G. sulfurreducens 对 U(Ⅵ) 的还原（Cologgi et al.，2011）。Jeon 等（2004）用多种天然 Fe(Ⅲ) 氧化物、合成 Fe(Ⅲ) 氧化物探究其对 G. sulfurreducens 还原 U(Ⅵ) 影响，结果发现，有天然 Fe(Ⅲ) 氧化物参与时，U(Ⅵ) 的还原速率大大降低了，这表明，在近中性环境中，水体中 U(Ⅵ) 可被还原去除，但在非还原性条件下，随时间增长，会发生吸附解除（Ortiz-Bernad et al.，2004）。

  Finneran 等（2002）在新墨西哥州的一个铀矿渣堆附近采集沉积物和水进行试验，发现在厌氧条件下添加醋酸或葡萄糖可以加速 U(Ⅵ) 还原，15d 内溶解性 U(Ⅵ) 浓度从 $10\mu mol/L$ 降到 $1\mu mol/L$ 以下。这一现象主要是由于微生物蛋白酶还原作用的结果，非生物还原剂 Fe(Ⅱ)、硫化物和 AQDS 的添加并没有对 U(Ⅵ) 的还原起到任何贡献。在原位生物修复中，以乙醇作为电子供体时 U(Ⅵ) 的还原速率比添加其他电子供体时要快（Wu et al.，2006）。同样针对科罗拉多州的一个铀矿加工厂的铀污染，Anderson 等（2003）侧重于含水层的 U(Ⅵ) 污染修复，对 15 个监测井的顺梯度和控制井的反梯度进行了评估。结果发现，连续 9d 添加醋酸后，某些井中水溶性 U(Ⅵ) 的浓度从 $0.6\sim1.2\mu mol/L$ 降低到 $0.18\mu mol/L$，而且

Fe(Ⅱ)含量呈现出增加趋势。但50d之后，U(Ⅵ)的浓度又开始增加，Fe(Ⅱ)含量也开始降低。这一实验说明，大部分生物质和地杆菌细胞对U(Ⅵ)污染修复与Fe(Ⅱ)的产生存在关联。Anderson等（2003）推测，地杆菌主要负责U(Ⅵ)和Fe(Ⅲ)的催化还原，而后来U(Ⅵ)浓度的再次升高是因为在实验点硫酸盐还原菌进行硫酸盐还原的活动占了主导作用。Vrionis等（2005）对U(Ⅵ)和Fe(Ⅲ)的催化还原效率最高的区域内生长的地杆菌进行16S rRNA基因序列分析，发现有一种菌的基因组序列编码与地杆菌 *Geobacter bemidjiensis* 最相符，并在实验前期占主导地位，而其他菌与地杆菌 *Geobacter lovleyi* 相符，但这些菌在实验后期占主导地位。这些实验现象说明了U(Ⅵ)原位生物修复的复杂性，并且也提示我们应该如何有效实施长时间的U(Ⅵ)原位生物修复。

1.2.3 电子穿梭体——微生物胞外呼吸的"催化剂"

电子穿梭体（electron shuttle）是具有氧化还原活性的化学物质，其主要功能是促进微生物胞外电子传递。一种优良的电子穿梭体应具备如下物理化学性质：a.电子穿梭体的氧化还原电位要在电子受体和电子供体的氧化还原电位之间；b.电子穿梭体具有较好的氧化还原特性，能够实现氧化态和还原态之间的快速转换；c.电子穿梭体结构稳定，且不参与细胞内的代谢过程；d.电子穿梭体不容易被细胞和电子受体吸附，具有一定的可溶性或导电性。电子穿梭体根据其来源不同可以分为人造电子穿梭体、天然电子穿梭体和内源性电子穿梭体。

1.2.3.1 人造电子穿梭体

人造电子穿梭体在生物电化学工程中的应用较多，常常在没有电活性微生物的条件下用来尝试降解复杂的底物。人造电子穿梭体降低了电子传递产生的过电势，并且能透过细菌生物膜获得电子。蒽醌-2,6-二磺酸盐（AQDS）是目前最常见的人造电子穿梭体，其常被作为电子转移机制研究中的模式电子穿梭体，它不仅能够促使微生物持续产电，而且对于重金属和有机污染物的微生物转化过程具有重要的促进作用（Van der Zee and Cervantes, 2009）。人造电子穿梭体在酵母微生物燃料电池中也是必不可少的，因为电子传递链位于细胞质线粒体中，而且通常情况下，像 *Saccharomices cerevisiae* 这样的酵母不能产生内源性电子穿梭体。在一些电化学电池中，微生物细胞很难维持细胞活性，但引入人造电子穿梭体甲基紫精（methyl viologen），可以有效地将电子传递至微生物，使这类微生物可以在溶液中进行呼吸作用。

1.2.3.2 天然电子穿梭体

天然电子穿梭体主要包括腐殖质、半胱氨酸、生物炭、S^0/HS^-电对、中性红、亚甲基蓝等。

半胱氨酸在自然界中较常见，它是一种含有巯基的氨基酸，是血红素蛋白、铁氧化还原蛋白和红素氧还蛋白等蛋白参与电子传递的必需氨基酸，也是一种微生物生长基质中普遍采用的还原剂。Kaden等（2002）发现，*Geobacter sulfurreducens*和*Wolinella succinogenes*共培养过程中半胱氨酸能够在种间传递电子。此外，在纯培养的*Geobacter sulfurreducens*中加入半胱氨酸可使胞外还原速率增加8~11倍。相比于普通的、可以形成未定Fe(Ⅲ)复合物的配位剂，半胱氨酸可以作为一种电子穿梭体增加微生物还原铁的速率（Doong and Schink，2002；Liu et al.，2012）。此外，半胱氨酸作为电子穿梭体还可以间接提高微生物燃料电池的效能。

生物炭是由生物残体在缺氧情况下经高温慢热解产生的一类富含芳香性和醌类结构的物质，其具有很强的氧化还原活性（Klüpfel et al.，2014），而且能够参与环境中许多非生物过程的氧化还原反应。基于电化学方法的研究表明，每克生物炭能够接受和供给数百摩尔电子，其具体数值因生物炭来源和炭化温度的不同而异（Klüpfel et al.，2014）。实际上，生物炭不仅能够参与和促进非生物的氧化还原过程，而且在微生物介导的氧化还原过程中也具有重要作用。Kappler等（2014）的研究表明，生物炭在5~10g/L的浓度下对于希瓦氏菌（*Shewanella oneidensis* MR-1）还原水铁矿的反应速率和反应程度均具有明显的促进作用，但过低的生物炭浓度（<1g/L）则对水铁矿的微生物还原起反作用，由此表明合适浓度的生物炭在微生物胞外电子传递过程中具有重要的电子穿梭体功能。进一步的控制实验发现，生物炭之所以具有很强的电子穿梭体的功能并不是因为生物炭中可溶性部分的有机结构在起作用，而与生物炭本身的微粒性质有关（Kappler et al.，2014）。然而，这个研究只是在实验室模拟中进行的，而在实际土壤等环境介质中，生物炭在矿物-微生物-生物炭联合形成的网络结构中是否还能够发挥以及能够多大程度发挥电子穿梭体的功能还有待于进一步探索。

硫是影响铁的生物地球化学氧化还原过程的重要因素，特别是还原态的硫与铁的反应活性更高（Lohmayer et al.，2014）。H_2S可导致Fe(Ⅲ)氧化物发生化学还原溶解（Afonso and Stumm，1992），并伴随着Fe(Ⅲ)氧化物表面吸附物质的释放。然而，H_2S并不一定是导致Fe(Ⅲ)氧化物还原必须存在的形态。很多硫还原细菌可以使用不同的氧化态硫作为电子受体，形成硫化物，从而促进 Fe(Ⅲ) 化学还原的发生和提高砷的移动性（Burton et al.，2013）。在硫代硫酸盐浓度较低时，硫还原细菌 *Sulfurospirillum deleyianum* 可以将甲酸盐氧化成 CO_2，并伴随着 Fe(Ⅲ) 还原成 Fe(Ⅱ)，由此推测氧化态硫与其相应的还原态硫构成的氧化还原电对可以很好充当电子穿梭体的角色。然而，目前还没有鉴别电子传递至 Fe(Ⅲ) 这一过程中氧化形成的硫化合物，但是它们可能是硫或者硫代硫酸盐。在碱性环境中，S^0/HS^- 电对作为电子穿梭体可以明显促进微生物还原 Fe(Ⅲ) 氧化物（Flynn et al.，2014），表明异化铁还原菌与硫的微生物还原过程存在密切联系（Friedrich and Finster，2014）。在三氯乙酸被还原为二氯乙酸的还原脱氯过程中，人们也评估了 S^0/S^{2-} 电对作为电子传递中介体所起的作用（de Wever et al.，2000）。另外，在缺氧呼吸过程中还发现了其他氧化还原电对可以作为电子穿梭体，如 CO_2/甲酸、$2H^+/H_2$、NO_3^-/NO_2^-、MnO_2/Mn^{2+}（Schröder，2007）。

1.2.3.3 内源性电子穿梭体

内源性电子穿梭体是微生物为了适应某种环境条件而分泌合成的有机化合物，目前已经发现的内源性电子穿梭体主要有黄素、吩嗪类物质、黑色素、苯醌类物质等。

(1) 黄素

黄素分子主要以黄素腺嘌呤二核苷酸（FAD）和黄素单核苷酸（FMN）两种形式存在，它们能够催化细菌内部氧化还原反应。其中核黄素是 FAD 和 FMN 主要的氧化还原部分（Tan et al.，2012）。von Canstein 等（2008）研究得出，很多 *Shewanella oneidensis* 能利用核黄素来介导不溶性三价铁氧化物还原。Marsili 等（2008）确证了 *Shewanella oneidensis* MR-1 和 *Shewanella oneidensis* MR-4 在序批式培养过程中能积累 250～500 nmol/L 的黄素，并且以这些黄素为可溶性电子传递中介体将胞外电子传递至处于氧化电势的电极。在 bfe 基因（负责 FAD 运输）缺陷型 *Shewanella oneidensis* MR-1 突变株的培养液中没有检测到黄素的存在，该突变株还原不溶性电子受体的能力也相应缺失，但向培养液中补充黄素后可以弥补这种还原能力的缺失；通过检测 Fe(Ⅱ) 含量和电流大小证明了黄素对电子转移能力的贡献率可达 75%，表明基于黄素介导的间接电子传递机制在整个电子转移过程中起到重要作用。*Shewanella oneidensis* MR-1 自身分泌的黄素作为外膜蛋白传导电子的氧化还原辅因子，提高了 OM c-Cyt 转移电子的能力（Okamoto et al.，

2013)。全细胞的差示脉冲伏安法显示黄素的氧化还原电位可逆地向正向移动了多于 100 mV，这与微生物产电增加的现象是相符的。更重要的是，试验结果表明，黄素/OM c-Cyt 之间的相互作用是加速了通过半醌发生的一电子（one-electron）氧化还原反应，且反应速率相比于黄素的单独作用快了 $10^3 \sim 10^5$ 倍。这个机理虽然与之前描述的氧化还原媒介参与的电子传递机制不同，但是黄素/OM c-Cyt 的相互作用同时调节了胞外电子传递的程度以及胞内代谢活动。

（2）吩嗪类物质

吩嗪类物质是 γ 变形细菌 Pseudomonas sp. 分泌的一类次生代谢物，功能基团是杂环结构，主要包括绿脓菌素、1-甲酰胺吩嗪、1-羟基吩嗪。吩嗪类物质的合成主要通过群体感应系统调节，群体感应系统通过调节 phzA-G 操纵子基因来合成 1-羟酸吩嗪，然后 1-羟酸吩嗪再合成其他吩嗪类分子。吩嗪分子具有与 AQDS 类似的功能，也可作为电子穿梭体。含氮杂环的吩嗪分子的电子传导能力取决于亲和的质子-电子传递过程，这与三元有机杂环的性质密切相关。但是，关于吩嗪如何参与多种电子传递过程的信息目前非常有限，尤其是吩嗪分子的吸电子基团与供电子基团特征，以及功能基团对电子转移的影响至今仍不清楚。杂环上取代基的不同是吩嗪分子物理和化学性质不同的主要原因，据报道取代基的特点和位置主要决定了吩嗪分子的氧化还原电势、极性和稳定性。

（3）黑色素

黑色素是在生物界普遍存在的一类羟基化聚合物，主要可分真黑色素、棕黑色素、异黑色素和脓褐素。该类物质具有的凝聚相物理性质使其具有导电性能，从而在微生物的电子转移中起到了电子穿梭体的作用。通过对比去除希瓦氏菌 algae BrY 细胞膜表面黑色素和没有去除黑色素的细菌，发现后者在还原含水氧化铁时的速率是前者的 10 倍，推测希瓦氏菌 MR-1 可有效利用环境中的酪氨酸合成脓褐素，从而通过促进固态金属氧化物的还原。

（4）苯醌类物质

苯醌除了被认为是天然有机物中的苯醌成分之外，许多微生物也能够产生这类化合物，如希瓦氏菌产生的甲基萘醌（Hernandez and Newman，2001；Newman and Kolter，2000），灰黄链霉菌的 Cinnaquinone 渗出物（Glaus et al.，1992），以及希瓦氏菌分泌的黑色素中的奎宁功能基团（Turick et al.，2002）。随着对可溶性电子传递中介体的关注，基于培养实验来鉴定内源介体的研究已经成为人们研究的热点。

1.2.4　土壤腐殖质——电子穿梭体的"聚合体"

1.2.4.1　土壤腐殖质的物质来源

土壤腐殖质的来源极为复杂，通常认为，植物残体是土壤腐殖质的主要来源，其次是微生物残体及其分泌物，土壤动物的贡献最小（Kögel-Knabner，2002）。在陆地生态系统中，进入土壤有机质的数量、化学组成及其化学结构特征对控制土壤腐殖质的形成与转化具有十分重要的作用（Kleber et al.，2007），不同来源的腐殖质在土壤中的命运是有差异的（Wiesenberg et al.，2008）。作为土壤腐殖质重要来源的植物残体在化学组成上与微生物残体、动物残体存在严重的不同，就植物残体本身而言，不同种类、器官、年龄的植物之间也存在显著区别（窦森，2010）。

以往一直认为微生物残体以及分泌物对土壤腐殖质的贡献不超过5％，然而，由于缺乏有效的分析手段可能会使得估算土壤中微生物来源的腐殖质的相对贡献存在很大的不确定性（Dalal，1998）。Clemmensen 等（2013）采用碳同位素示踪技术和分子技术对北方森林土壤碳的来源问题进行了评估，发现土壤中50％~70％的碳储量主要是来自根系及与其共生的菌根真菌，尤其是在土壤根系密度较大的层位更为明显，这显然颠覆了以往一直认为的土壤腐殖质主要是来自地上部分植物残体凋落物的认识。传统上认为微生物残体以及微生物分泌物不是构成腐殖质的主要前体物，然而，基于最近对腐殖物质产生的新认识（Sutton and Sposito，2005），发现腐殖质是植物残体和微生物残体分解的中间产物的混合物质（Kelleher and Simpson，2006）。Simpson 等（2007）通过对土壤有机质进行 NMR 分析，发现土壤中45％的腐殖质是微生物残体或微生物分泌物贡献的。由于黑色素具有与腐殖质相类似的结构组成，因此其通常被推测为腐殖质形成的前体物（Saiz-Jimenez，1996）。然而目前关于土壤中黑色素含量以及分解行为的研究较少，这主要因为黑色素的提取方法和分析技术还存在一定的困难。

1.2.4.2　土壤腐殖质的组成特征

腐殖质在化学上是不均匀的多功能有机分子，在土壤中主要与矿物通过相互作用交织在一起，其含有大量的活性官能团，不仅具有络合金属离子的能力，而且具有很强的氧化还原特性（Aiken et al.，1985；Stevenson，1994）。

由于来源和成岩作用的差异,不同土壤样品腐殖质的化学成分及结构会存在很大的不同,因此无法利用一个统一的化学分子式来准确地对腐殖质进行描述。土壤腐殖质是由复杂的芳香及脂肪结构组成的大分子,这个大分子含有羟基、羧基、氨基、酚基等结构,可以通过红外和核磁共振光谱确定这些结构(Stevenson, 1994)。由于其化学结构的多变性,天然腐殖质一直都是按照其溶解性与分子量的标准进行划分。腐殖质可以分为胡敏酸(humic acid, HA,也称腐殖酸)、富里酸(fulvic acid, FA)和胡敏素(humin, HM),其中胡敏素具有比较高的分子量($20\sim100$kDa,1kDa=10^3g/mol,下同),只溶于碱性环境下,富里酸具有比较低的分子量($0.5\sim5$kDa),在所有的pH值条件下都可以溶解,胡敏素占的比例最大,在所有pH值下都无法溶解(Aiken et al., 1985; Stevenson, 1994)。

与通过大分子结构来描述腐殖质特性不同,Piccolo(2001)、Sutton和Sposito(2005)以及Kleber和Johnson(2010)提出了一种新的概念模型来描述土壤腐殖质的结构。根据该模型,土壤腐殖质并不是由单个的大分子组成的,而是由各种有机小分子团通过疏水作用及氢键结合在一起形成的聚合体。当地球化学条件发生变化时,这些键会被加强或者减弱,进而引起结构性的变化,甚至会导致单个有机小分子与聚合体分离(Kleber and Johnson, 2010; Sutton and Sposito, 2005)。对腐殖质有贡献的可以是各种类型的有机分子,例如脂肪酸、羧酸、乙醇、木质素、糖等,按照原有的腐殖质大分子结构的定义,这些有机分子甚至有可能都不是腐殖质的组成部分(Stevenson, 1994)。然而,新的概念模型将这些有机分子作为腐殖质的组成部分(Kleber and Johnson, 2010; Sutton and Sposito, 2005)。

1.2.4.3 土壤腐殖质的形成机制

腐殖质的形成至今仍是一个谜,因为要研究探索的问题太重要、太复杂以至于出现腐殖质学或腐殖质研究的说法(Paul, 2002)。目前对土壤腐殖质的形成机制有多糖-酰胺缩合学说、煤化学说、木质素学说、木质素多酚学说、微生物多酚学说、微生物合成学说、细胞自溶学说(包括植物和微生物)和厌氧发酵学说等多种学说(Stevenson et al., 1994;李阜隶,2003;窦森,2010),其中微生物合成学说、微生物多酚学说、厌氧发酵学说强调微生物的作用,木质素学说强调植物的作用,木质素多酚学说和细胞自溶学说同时承认微生物和植物的作用,煤化学说和糖-酰胺缩合学说则强调纯化学反应的作用。在上述观点中,核心问题是微生物是否参与或在哪个环节参与腐殖质的形成。不过从土壤学的角度,这种分歧的焦点更加具体,即微生物是否直接合成腐殖质或其前体成分。对此持肯定观点的是多酚学说,对此持否定观点的是木质素学说。在解释腐殖质形成途径时,通常认为几种途径可能在各种土壤中都存在,只不过各自所占比重不同。对腐殖质生成机理认识的差异,也影响到对胡敏酸和富里酸形成顺序的推断,按木质素理论是先形成胡敏酸,然后由胡敏酸裂解成富里酸;按多酚理论是先形成富里酸,再由富里酸聚合成胡敏酸,尽管也存在着直接形成胡敏酸的可能性,这意味着在对待腐殖质各组分的演化方向的问题上也存在着根本对立的观点。此外,还有研究显示,不同环境条件下胡敏酸和富里酸的形成顺序也存在差异(Stevenson et al., 1982;窦森,2010)。目前我们只了解了腐殖化过程的一般轮廓,对于过程中的很多具体内容还有待于进一步的研究。

1.2.4.4 土壤腐殖质与类腐殖质的界定

以往很多关于腐殖质的研究有一个特点,像对待土壤一样,把新鲜有机物料、有机肥或菌体本身,以及加入这些物质的土壤作为供试材料,把它们的碱提取液当成胡敏酸和富里酸,剩下的残余物当成胡敏素。但事实上,这些提取物或残余物未必就是真正的腐殖质(Bottomley,1915;熊田恭一,1981),而有些只不过是胡敏酸和富里酸的原始形态(科诺诺娃,1959),因此,我们不能仅仅根据暗色和溶解性来判断提取物是否就是腐殖质,因此明确概念、统一方法和选择判断标准是研究土壤以外介质发生的暗色化现象与腐殖质生成机理的前提。人们对获得的类腐殖质与土壤腐殖质的异同点进行了对比(熊田恭一,1981),但却很少进一步论述异同点的判定标准。综合各种指标是区分类腐殖质与腐殖质的主要手段,过去仅从外观颜色或吸收光谱暗色物质定量来比较,后来用色调系数、相对色度、分解率、褪色率、E_4/E_6等,目前有条件使用更多的结构特征指标,如元素组成、数均分子量、官能团含量、活化度、红外光谱、荧光光谱、紫外光谱、热稳定性、核磁共振波谱、电子显微镜观察等。

1.2.4.5 土壤腐殖质的分解与转化过程

细菌虽然在自然界中分布很广泛,且可以参与腐殖物质的降解,但其降解能力十分有限(Dehorter et al., 1992;Esham et al., 2000;Filip and Tesarova, 2004)。相比而言,真菌具有较强的降解腐殖物质的能力(Qi et al., 2004),特别是担子菌中的白腐真菌和凋落物降解菌。子囊菌降解腐殖物质的能力虽然不及白腐真菌和凋落物降解菌,但它对促进土壤中腐殖物质的降解与转化方面仍有相当大的贡献（Rezacova et al.，2006）。白腐真菌和凋落物降解菌是降解木质素的主要菌类（Hatakka，1994），然而，人们发现它们也能有效降解比木质素的结构更为复杂的腐殖物质（Hofrichter and Fritsche，1996；Gramss et al.，1999），这主要是由于它们所产生的非特异性的过氧化物酶（LiP）、锰过氧化氢酶（MnP）与漆酶（EC）能够与腐殖物质上的芳香族基团发生反应。另外，其他种类的木质素降解酶，如GLX、纤维二糖脱氢酶（cellobiose dehydrogenase，CDH）、AAO、多功能过氧化物酶（versatile peroxidase，VP）以及细胞色素P450对腐殖物质的降解与转化也起着十分重要的作用（Grinhut et al.，2007）。

白腐真菌和凋落物降解菌可以采取直接攻击和介导性攻击来降解与转化腐殖质（Grinhut et al.，2007），然而关于其具体的机理过程以及降解产物的物理与化学特征还有待于进一步研究（Grinhut et al.，2007；Grinhut et al.，2011）。腐殖质的降解机理是一个复杂的过程，它与真菌种类、生态环境以及底物特征密切相关（Grinhut et al.，2007）。Dehorter等（1992）将来源于森林土壤的胡敏酸作为基底物质进行微生物培养试验，发现尽管胡敏酸减少了30%，但^{13}C-核磁共振（^{13}C-NMR）的结果却显示胡敏酸的结构特征没有发生明显的变化。然而，在以褐煤的胡敏酸为基底物质来培养白腐真菌的试验中，却发现随着胡敏酸的降解羧基官能团以及脂肪族碳链上的羟基与甲氧基官能团增多，而芳香基官能团则减少（Dehorter et al.，1992；Willmann and Fakoussa，1997）。

1.2.4.6 土壤腐殖质的稳定机制

（1）对分子结构决定论的新认识

将土壤腐殖质视为难降解有机物的观点已经流行了很长一段时间（Stevenson，1994）。以往研究认为腐殖质是生物过程中自发产生的有机物质（Guggenberger，2005），因此它的分子大小与结构具有高度多样性，进而导致土壤中没有相应的特异性降解酶能够对其进行降解（Hedges，1988；Stevenson，1994）。然而，有些研究却认为腐殖物质的结构发生微小的变化都会导致微生物酶发生相应的变化，从而导致腐殖物质的分解成为可能（Sutton and Sposito，2005）。另外，Kelleher和Simpson（2006）通过NMR技术发现腐殖物质并不是一种化学特异性的有机物，而是一种由微生物和植物的生物聚合物结合而成的复杂混合物，而且生物聚合物之间结合力可能也仅是微弱的分子间作用力，例如疏水键和氢键（Sutton and Sposito，2005），这也表明了腐殖物质并非一种难以降解的有机物。

以往一直认为腐殖物质是土壤中稳定碳库的主要组成部分，然而，许多学者通过原位直接观测技术却发现腐殖物质在土壤总有机质中仅占很小的一部分（von Lützow et al.，2006；Kleber and Johnson，2010），而用化学方法从土壤中提取出的胡敏酸类化合物可能是由燃烧产生的（Trompowsky et al.，2005），这对腐殖物质的形成是稳定土壤碳库的重要机制也提出了质疑。最新研究表明，土壤腐殖质稳定与否主要是受微环境系统影响，而并不是由分子结构来决定（Schmidt et al.，2011），这一观点对我们重新认识腐殖质在地球环境中的作用与意义提供了新的思路。

（2）物理保护机制

土壤腐殖质的物理保护机制主要是通过将腐殖质包裹在土壤团聚体与微小孔隙中来避免微生物的接近与攻击（Sollins et al.，1996；Six et al.，2000）。尽管土壤中 Fe、Al 氧化物或氢氧化物通过非生物因素形成的网状结构是一种重要的腐殖质包裹体（Mayer et al.，2004），但包裹腐殖质主要还是通过生物因素形成的土壤团聚体来实现（Oades，1993；Six et al.，2002）。土壤微团聚体相对于大团聚体具有更好的保护土壤腐殖质的功能（Six et al.，2004）。然而，许多研究发现，相对于粗粒组分，富含微团聚体的粉砂和黏粒组分具有更高的营养利用效率和更强的抵抗捕食者入侵的能力，因此具有更高的微生物多样性（Selesi et al.，2007），这表明仅仅通过微团聚体具有排斥微生物攻击的特性是不足以解释土壤微团聚体是稳定腐殖质的重要场所的观点（McCarthy et al.，2008）。计算机微断层摄像技术是一种非破坏性观察土壤微型结构的十分有利的手段，Peth 等（2008）通过该技术观察到土壤微团聚体中具有高度相互连接的微孔网状结构。这种微孔网状结构可以允许氧气以及土壤溶液在其中自由传输，但其中的腐殖质由于被孔隙内壁所吸附或与矿物相互作用不能任意移动，因此，尽管土壤微团聚体中具有丰富的微生物种类，但由于微生物所分泌的酶被固定而无法发挥其应有的生理活性，从而使得微团聚体中的腐殖质能够免受酶的接近而得到很好的保存。然而，目前这方面的研究还仅仅处于初步阶段，关于土壤微团聚体的生物多样性、酶活性以及微孔网状结构之间的具体关系还有待于进一步探索。

土壤腐殖质嵌入层状硅酸盐黏土矿物的夹层中也是一种很好抑制微生物攻击的方式（Kennedy et al.，2002）。尽管土壤中大分子腐殖质在结构上似乎要大于黏土矿物的夹层，但是某些腐殖质的可溶性与弯曲性可使其能够很好地嵌入黏土矿物的夹层中（Schnitzer et al.，1988）。研究表明，嵌入蒙脱石夹层中的有机小分子化合物可发生构象变化，并且在 Al^{3+} 或 Fe^{3+} 的催化作用下可以进一步聚合成大分子的腐殖质（Kennedy et al.，2002），这对进一步加强腐殖质的稳定性起到了十分重要的作用。黏土矿物夹层中腐殖质的稳定性可能与同构置换有密切的关系，Tunega 等（2007）研究发现蛭石与云母在层状硅酸盐结构的外表层具有丰富的同构置换，从而使其能够形成十分稳定的层间有机分子。迄今为止，对黏土矿物夹层中腐殖质的分析仍缺乏十分有效的手段，对它们的结构与定量研究尚未有清晰的认识（Eusterhues et al.，2003），因此关于黏土矿物夹层腐殖质的稳定机制研究仍需要进一步探索。

（3）化学保护机制

土壤腐殖质与土壤矿物通过相互作用形成腐殖质-矿物复合体是防止土壤腐殖质分解的重要途径（Kalbitz et al.，2005），其相互作用的方式主要有静电作用（阴离子交换）、配位反应、疏水作用、范德华力、氢键、阳离子桥键、水桥以及熵变驱动的物理吸附等（Mikutta et al.，2007）。不同类型的矿物对土壤腐殖质的吸附机制不同，例如，Fe、Al 氧化物矿物以配位反应为主，蒙脱石以阳离子桥键为主，而高岭石则以范德华力为主，因此，根据这些结合力的大小，Fe、Al 氧化物矿物所结合的土壤腐殖质具有最强的抵抗微生物降解的能力，蒙脱石次之，而高岭石最弱（Mikutta et al.，2007）。通常情况下，土壤中腐殖质-矿物复合体的含量与 Fe 氧化物的含量呈正相关关系（Mikutta et al.，2006）。

许多研究认为，土壤腐殖质是以单层膜的形式连续分布在土壤矿物的表面，由此指出土壤矿物表面积大小可以用于指示土壤中固持腐殖质含量，即大的土壤矿物表面积有利于固持更多的土壤腐殖质（Wagai et al.，2009）。然而，也有许多报道称土壤腐殖质是以斑块的形式吸附在土壤矿物表面，指出土壤矿物表面积与土壤腐殖质的固持之间没有必然的联系（Kögel-Knabner et al.，2008）。由此可见，关于土壤矿物表面积与腐殖质的固持之间的确切关系仍需要进一步探索。

土壤腐殖质与金属离子相互作用形成络合物也是稳定腐殖质的重要机制，金属离子对微生物的毒性作用以及金属离子对胞外酶的钝化作用可能是造成有机金属络合物具有更高稳定性的重要原因（von Lützow et al.，2006）。另外，络合作用导致的腐殖质品质、分子大小、电荷以及空间结构的变化也可能对稳定腐殖质有一定的贡献（McKeague et al.，1986）。此外，腐殖质与金属离子形成络合物沉淀在一定程度上也可以很好地避免微生物的攻击，特别是对于大分子腐殖质更为普遍（Schwesig et al.，2003）。参与络合作用的金属离子主要有 Ca^{2+}、Al^{3+} 和 Fe^{3+}（Baldock and Skjemstad，2000）。许多研究都一致认为土壤腐殖质通过与 Al^{3+}、

Fe^{3+}相互作用能够有效地提高其抗降解能力（Nierop et al.，2002）。然而，目前大部分研究结论都是基于间接的证据，对于金属离子对腐殖质稳定性的影响还难以定量描述。

1.2.4.7　土壤腐殖质的电子穿梭官能团

土壤腐殖质同时具有接受电子和提供电子的功能，说明腐殖质可以参与氧化还原反应，不仅是酸碱可溶的胡敏酸和富里酸具有电子转移能力，而且固相的酸碱不溶物胡敏素也具有一定的电子转移能力（Zhang and Katayama，2012）。土壤腐殖质之所以具有很强的电子转移能力主要是由于其结构中含有丰富的电子穿梭官能团。之前很多开创性的工作都是基于腐殖质对硝基芳香化合物还原的影响研究，研究结果显示，腐殖质中奎宁能够刺激氧化还原过程的电子转移（Dunnivant et al.，1992；Tratnyek and Macalady，1989），说明奎宁具有电子穿梭功能。后来有很多证据都支持了这样一个假设，那就是醌类是腐殖质中主要的电子穿梭官能团。电子自旋共振波谱分析直接证明了腐殖质中的醌基团是微生物还原过程中真正接受电子的官能团（Scott et al.，1998）。此外，基因的研究结果显示，*Shewanella oneidensis* MR-1对奎宁和腐殖质的还原都是基于一个共同的生物化学过程（Newman and Kolter，2000），说明腐殖质确实存在与奎宁相类似的官能团。进一步的研究表明，在腐殖质和AQDS的微生物还原过程中，甲基萘醌是*Shewanella oneidensis* MR-1电子传递链的必不可少的物质，而缺乏合成甲基萘醌能力的突变体则无法还原AQDS与腐殖质（Newman and Kolter，2000）。通常情况下，腐殖质中的醌含量与腐殖质的电子转移能力存在很好的一致性（Sposito，2011），这也说明了醌基团对腐殖质电子转移具有很大的贡献。还有报道显示，从不同富含有机物环境介质中提取的各种胡敏酸，其电子转移能力与其对应于醌基团的红外光谱强度之间有着很强的联系（Hernández-Montoya et al.，2012），这间接证实了腐殖质中存在醌类的电子穿梭官能团。此外，Aeschbacher等（2010）采用电化学的方法对购买的各种腐殖质标准品进行了电子接受能力的测定，发现腐殖质标准品的电子接受能力与C/H摩尔比及芳香性之间都存在着线性关系，暗示着醌基团可能是影响这些氧化还原反应的主要因子，这与Ratasuk和Nanny（2007）所报道的结果相一致。综上所述，醌基团是腐殖质中非常重要的一类电子穿梭官能团。

实际上，土壤腐殖质中的非醌官能团对其电子转移能力也会起到显著的贡献。采用H_2/Pd反应体系和pH=8的实验条件对腐殖质进行还原后，发现腐殖质在红外光谱1360cm^{-1}处会表现出与对苯二酚基团有关的光谱信号；相反地，采用

H_2/Pd 反应体系和 pH=6.5 的实验条件对腐殖质进行还原后，却发现腐殖质并没有表现出对苯二酚的红外光谱信号，但其芳香酮基团却没有什么变化（Hernández-Montoya et al.，2012）。上述现象表明，在 H_2/Pd 反应系统中，pH=6.5 时可以很好地规避腐殖质中醌基团的电子穿梭功能的有效性，这主要是由于在这种体系下醌会被质子化形成酚基（pK_a=9.9），从而阻碍了醌基团的电子转移。Ratasuk 和 Nanny（2007）的研究表明在 pH=6.5 的情况下，采用 H_2/Pd 反应系统对胡敏酸进行还原后，结果发现这种胡敏酸并没有失去电子转移能力，仍然能够促进柠檬酸铁的还原，说明腐殖质中还存在其他非醌类的电子穿梭官能团。基于类似这种规避醌基团电子穿梭功能有效性的方法，人们对来自各种富含有机物的环境介质中的腐殖质样品进行了分析，结果显示，腐殖质中非醌类官能团占总电子转移能力的 44%~58%（Hernández-Montoya et al.，2012；Ratasuk and Nanny，2007）。进一步的研究表明，在 pH=6.5 时，H_2/Pd 反应系统对规避醌基团电子穿梭功能的机理是可逆的，许多微生物（例如地杆菌 *G. sulfurreducens*）能够同时还原各种腐殖质中的醌类及非醌类官能团（Hernández-Montoya et al.，2012）。在这些腐殖质样品中检测到的具有明显电子转移能力的非醌官能团可能与含氮及含硫氧官能团有关，例如二甲基砜、3-甲硫基丙酸、*n*-甲基苯胺、1-甲基-2,5-吡咯烷二酮等（Fimmen et al.，2007）。此外，还有研究显示，腐殖质中络合金属的氧化还原电对也会对腐殖质的电子穿梭功能起到一定的贡献作用（Sposito and Struyk，2001），但这种作用可能相对较小。

目前很多相关学者为了更深入理解腐殖质在非生物的与微生物介导的电子转移过程中所涉及的化学结构及氧化还原性质进行不懈努力。很多光谱与色谱分析技术都可以用于表征腐殖质的化学结构，其中最常用的非破坏性分析方法包括红外光谱、紫外可见光谱、荧光光谱、核磁共振波谱、质谱及其相关技术、X射线技术、色谱层析技术（如凝胶电泳）等。然而，迄今为止，最有用的分析腐殖质氧化还原性质的方法主要是电子顺磁共振波谱、X射线吸收近边结构光谱、傅里叶转换红外光谱和荧光光谱，这些方法对揭示腐殖质的氧化还原状态和氧化还原活性已经做出了许多贡献。氧化还原状态指的是氧化还原活性官能团被氧化或被还原的相对程度；而氧化还原活性则是指腐殖质接受或者提供电子的能力。尽管目前关于腐殖质氧化还原状态和氧化还原活性的化学层面的研究取得了很大的进步，但是依然存在许多问题有待于进一步研究，以便更全面地理解和预测自然生态系统中腐殖质的生物地球化学行为以及工程系统中由腐殖质介导的氧化还原过程。

1.3 研究思路与研究内容

1.3.1 研究目标

评估土壤原位固相腐殖质的电子转移能力，明确影响土壤原位固相腐殖质电子转移能力的关键性因素，揭示土壤腐殖质在生物地球化学的氧化还原过程中作为胞外电子穿梭体的持续能力，阐明土壤腐殖质电子转移能力对气候变暖和土地利用变化的响应机制，以期为深入认识在人类活动背景下土壤腐殖质电子转移机制及其调控污染物环境行为的机理提供理论基础。

1.3.2 研究内容

(1) 土壤固相腐殖质的电子转移能力研究

联合微生物还原法、电化学方法与交替式的氧化还原方法，分析土壤固相腐殖质的电子接受能力、电子供给能力与电子循环能力，研究土壤固相腐殖质的电子转移能力，并通过与土壤溶解性腐殖质的电子传递能力进行比较，评估土壤固相腐殖质作为胞外电子穿梭体的能力。

(2) 土壤溶解性腐殖质的电子循环能力研究

通过设计微生物还原和氧气重新氧化的循环周期实验，模拟土壤实际存在的间歇性缺氧和曝气的循环过程，联合微生物还原法与电化学方法，研究土壤腐殖质作为胞外电子穿梭体的持续能力；基于化学氧化还原循环实验，分析土壤腐殖质中苯醌基团和非苯醌基团的电子承载能力，阐明不同基团对土壤腐殖质电子转移持续能力的相对贡献。

(3) 土壤腐殖质电子转移能力对增温的响应

通过采集中国黄土剖面冰期-间冰期的土壤集、中国等400mm降雨量线不同纬度梯度的土壤集与中国东灵山不同海拔梯度的土壤集，提取土壤中的胡敏酸与富里酸，采用电化学方法对二者的电子接受能力与电子供给能力进行分析，并通过结合化学结构表征手段和土壤环境因子的变化，研究土壤腐殖质电子转移能力对温度增加的响应。

(4) 土壤腐殖质竞争性抑制甲烷生成对增温的响应

在不同温度条件下，对添加不同腐殖质的水稻田土壤和湿地土壤进行厌氧培养，监测甲烷生成的动态过程，验证土壤腐殖质的微生物还原过程对甲烷生成的竞争抑制作用，阐明这种抑制作用对温度增加的响应机制。

（5）土壤腐殖质电子转移能力对土地利用的响应

通过对不同农用地的土壤进行了采样，使用微生物还原法对土壤中胡敏酸和富里酸的电子转移能力进行量化，以评估土壤胡敏酸和富里酸电子转移能力对农用地类型的响应，同时研究土壤腐殖质电子转移能力对水稻田耕作年限的响应。

参考文献

Ackerley D F, Gonzalez C F, Keyhan M, et al, 2004. Mechanism of chromatereduction by the Escherichia coli protein, NfsA, and the role of different chromate reductasesin minimizing oxidative stress during chromate reduction. Environmental Microbiology, 6: 851-860.

Aeschbacher M, Sander M, Schwarzenbach R P, 2010. Novel electrochemical approach to assess the redox properties of humic substances. Environmental Science & Technology, 44 (1): 87-93.

Afkar E, Reguera G, Schiffer M, et al, 2005. Novel Geobacteraceae-specific outer membrane protein J (OmpJ) is essential for electron transport to Fe(III) and Mn(IV) oxides in Geobacter sulfurreducens. BMC Microbiology, 5: 41.

Afonso M D, Stumm W, 1992. Reductive dissolution of iron (III)(hydr) oxides by hydrogen-sulfide. Langmuir, 8: 1671-1675.

Aiken G R, McKnight D M, Wershaw R L, et al, 1985. Humic substances in soil, sediment and water. New York: A wiley Interscience publication.

Amonette J E, Workman D J, Kennedy D W, et al, 2000. Dechlorination of carbon tetrachloride by Fe(II) associated with goethite. Environmental Science & Technology, 34: 4606-4613.

Amstaetter K, Borch T, Larese-Casanova P, et al, 2010. Redox transformation of arsenic by Fe(II)-activated goethite (a-FeOOH). Environmental Science & Technology, 44: 102-108.

Anderson R T, Vrionis H A, Ortiz-Bernad I, et al, 2003. Stimulating the in situactivity of Geobacter species to remove uranium from the groundwater of a uraniumcontaminatedaquifer. Applied and Environmental Microbiology, 69: 5884-5891.

Arnold R G, Dichristina T J, Hoffmann M R, 1988. Reductive dissolution of Fe(III) oxides by *Pseudomonas* sp. 200. Biotechnology & Bioengineering, 32 (9): 1081-1096.

Bader J L, Gonzalez G, Goodell P C, et al, 1999. Aerobic reduction of hexavalentchromium in soil by indigenous microorganisms. Bioremediation Journal, 3: 201-211.

Baldock J A, Skjemstad J O, 2000. Role of the soil matrix and minerals in protecting natural organic materials against biological attack. Organic Geochemistry, 31: 697-710.

Barak Y, Ackerley D F, Dodge C J, et al, 2006. Analysis of novel soluble chromate and uranyl reductases and generation of an improved enzyme bydirected evolution. Applied and Environmental Microbiology, 72: 7074-7082.

Barkay T, Miller S M, Summers A O, 2003. Bacterial mercury resistance from atoms toecosystems. FEMS Microbiology Reviews, 27: 355-384.

Barkay T, Wagner-Dobler I, Allen I, et al, 2005, Microbial transformations of mercury: potentials, challenges, and achievements in controlling mercury toxicity in the environment. Advances in Applied Microbiology, 57: 1-52.

Barlett M, Moon H S, Peacock A A, et al, 2012a. Uranium reduction and microbial community development in response to stimulation with different electron donors. Biodegradation, 23: 535-546.

Barlett M, Zhuang K, Mahadevan R, et al, 2012b. Integrative analysis of Geobacter spp. and sulfate-reducing bacteria during uranium bioremediation. Biogeosciences, 9: 1033-1040.

Barrera-Díaz C E, Lugo-Lugo V, Bilyeu B, 2012. A review of chemical, electrochemical and biological methods for aqueous Cr(Ⅵ) reduction. Journal of Hazardous Materials, 223-224: 1-12.

Bauer I, Kappler A, 2009. Rates and extent of reduction of Fe(Ⅲ) compound sand O_2 by humic substances. Environmental Science & Technology, 43: 4902-4908.

Becerra-Castro C, Lopes A R, Vaz-Moreira I, et al, 2015. Wastewater reuse in irrigation: A microbiological perspective on implications in soil fertility and human and environmental health. Environment International, 75: 117-135.

Belchik S M, Kennedy D W, Dohnalkova A C, et al, 2011. Extracellular reduction of hexavalent chromium by cytochromesMtrC and OmcA of Shewanella oneidensis MR-1. Applied and Environmental Microbiology, 77: 4035-4041.

Beliaev A S, Saffarini D A, Mclaughlin J L, et al, 2001. MtrC, an outer membrane decahaem c-cytochrome required for metal reduction in Shewanella putrefaciens MR-1. Molecular Microbiology, 39 (3): 722-730.

Beller H R, 2005. Anaerobic, nitrate-dependent oxidation of U(Ⅳ) oxide minerals by the chemolithoautotrophic bacterium Thiobacillus denitrificans. Applied and Environmental Microbiology, 71: 2170-2174.

Benz M, Schink B, Brune A, 1998. Humic acid reduction by Propionibacterium freudenreichii and other fermenting bacteria. Applied and Environmental Microbiology, 64: 4507-4512.

Berner R A, 1999. A new look at the long-term carbon cycle. GSA Today, 9: 1-6.

Borch T, Inskeep W P, Harwood J A, et al, 2005. Impact of ferrihydrite and anthraquinone-2,6-disulfonate on the reductive transformation of 2,4,6-trinitrotoluene by a gram-positive fermenting bacterium. Environmental Science & Technology, 39: 7126-7133.

Borch T, Kretzschmar R, Kappler A, et al, 2010. Biogeochemical redox processes and their impact on contaminant dynamics. Environmental Science & Technology, 44: 15-23.

Borch T, Masue Y, Kukkadapu R K, et al, 2007. Phosphateimposed limitations on biological reduction and alteration offerrihydrite. Environmental Science & Technology, 41: 166-172.

Bottomley W B, 1915. The formation of humic bodies from organic substances, Biochem Great Britain, 9: 260-270.

Breuer M, Rosso K M, Blumberger J, 2014. Electron flow in multiheme bacterial cytochromes is a balancing act between heme electronic interaction and redox potentials. Proceedings of the National Academy of Sciences, 111: 611-616.

Brim H, McFarlan S C, Fredrickson J K, et al, 2000. Engineering deinococcus radiodurans for metal remediation in radioactive mixed wasteenvironments. Nature Biotechnologynology, 18: 85-90.

Bücking C, Popp R, Kerzemnacher S, et al, 2010. Involvement and specificity of Shewanella oneidensis outer membrane cytochromes in the reduction of soluble and solid-phase terminal electron acceptors. FEMS Microbiology Letters, 306: 144-151.

Burdgett R D, van der Putten W H, 2014. Belowground biodiversity and ecosystem functioning. Nature, 515: 505-511.

Burton E D, Johnston S G, Planer-Friedrich B, 2013. Sulfate availability drives divergent evolution of arsenic speciation during microbially mediated reductive transformation of schwertmannite. Environmental Science & Technology, 47: 2221−2229.

Butler J, He Q, Nevin K, et al, 2007. Genomic and microarray analysis of aromatics degradation in Geobacter metallireducens and comparison to a Geobacter isolatefrom a contaminated field site. BMC Genomics, 8: 180.

Butler J E, Young N D, Lovley D R, 2010. Evolution of electron transfer out of the cell: comparative genomics of six Geobacter genomes. BMC Genomics, 11: 40.

Canfield D E, 1989. Reactive iron in marine sediments. Geochimica et Cosmochimica Acta, 53: 619-632.

Cao B, Shi L, Brown R N, et al, 2011. Extracellular polymeric substances from Shewanella sp. HRCR-1 biofilms: characterization by infrared spectroscopy and proteomics. Environmental Microbiology, 13 (4), 1018-1031.

Carmona M, Díaz E, 2005. Iron-reducing bacteria unravel novel strategies for the anaerobic catabolism of aromatic compounds. Molecular Microbiology, 58: 1210-1215.

Carmona M, Zamarro M A T, Blázquez B, et al, 2009. Anaerobic catabolism of aromatic compounds: agenetic and genomic view. Microbiology and Molecar Biology Reviews, 73: 71-133.

Cervantes C, Campos-García J, Nies D, et al, 2007. Reduction and efflux of chromate by bacteria molecular microbiology of heavy metals. Springer, 407-419

Cervantes F J, Bok F A M, de Duong-Dac T, et al, 2002. Reduction of humic substances by halorespiring, sulphate-reducing and methanogenic microorganisms. Environmental Microbiology, 4: 51-57.

Charlet L, Silvester E, Liger E, 1998. N-compound reduction and actinide immobilisation insurficial fluids by Fe(Ⅱ): the surface Fe$_{Ⅲ}$OFe$_{Ⅱ}$OH species, as major reductant. Chemical Geology, 151: 85-93.

Chen J M, Hao O J, 1998. Microbial chromium (Ⅵ) reduction. Critical Reviews in Environmental Science and Technology, 28: 219-251.

Childers S E, Cifo S, Lovley D R, 2002. Geobacter metallireducens accesses insoluble Fe(Ⅲ) oxide by chemotaxis. Nature, 416: 767-769.

Chirwa E M N, Wang Y T, 1997. Hexavalent chromium reduction by Bacillus sp. in a packed-bed bioreactor. Environmental Science & Technology, 31: 1446-1451.

Christensen T H, Kjeldsen P, Bjerg P L, et al, 2001. Biogeochemistry of landfill leachate plumes. Applied Geochemistry, 16: 659-718.

Clark D L, Hobart D E, Neu M P, 1995. Actinide carbonate complexes and their importance in actinide environmental chemistry. Chemical Reviews, 95: 25-48.

Clarke T A, Cole J A, Richardson D J, et al, 2007. The crystal structure of the pentahaem c-type cytochrome NrfB and characterization of its solution-state interaction with the pentahaem nitrite reductase NrfA. Biochemistry Journal, 406: 19-30.

Clarke T A, Edwards M J, Gates A J, et al, 2011. Structure of a bacterial cell surface decaheme electron conduit. Proceedings of the National Academy of Sciences, 108: 9384-9389.

Clarkson T W, 1997. The toxicology of mercury. Critical Reviews in Clinical Laboratory Sciences, 34: 369-403.

Clemmensen K E, Bahr A, Ovaskainen O, et al, 2013. Roots and associated fungi drive long-term carbon sequestration in boreal forest. Science, 339: 1615-1618.

Coates J D, Bhupathiraju V K, Achenbach L A, et al, 2001. Geobacter hydrogenophilus, Geobacter chapellei and Geobacter grbiciae, three new, strictly anaerobic, dissimilatory Fe(Ⅲ)-reducers. International Journal of Systematic and Evolutionary Microbiology, 51: 581-588.

Coates J D, EllisD J, Gaw C V, et al, 1999. Geothrix fermentans gen. nov., sp. Nov., a novel Fe(Ⅲ)-reducing bacterium from a hydrocarbon-contaminated aquifer. International Journal of Systematic Bacteriology, 49: 1615-1622.

Coates J D, Phillips E J, Lonergan D J, et al, 1996. Isolation of Geobacter species from diverse sedimentary environments. Applied and Environmental Microbiology, 62: 1531-1536.

Coby A J, Picardal F, Shelobolina E, et al, 2011. Repeated anaerobic microbial redox cycling of iron. Applied and Environmental Microbiology, 77: 6036-6042.

Cologgi D L, Lampa-Pastirk S, Speers A M, et al, 2011. Extracellular reduction of uranium via Geobacter conductive pili as a protective cellular mechanism. Proceedings of the National Academy of Sciences, 108: 15248-15252.

Coursolle D, Gralnick J A, 2010. Modularity of the Mtr respiratory pathway of Shewanella oneidensis strain MR-1. Molecular Microbiology, 77 (4): 14.

Cutting R S, Coker V S, Fellowes J W, et al, 2009. Mineralogical and morphological constraints on the reduction of Fe(III) minerals by Geobacter sulfurreducens. Geochimica et Cosmochimica Acta, 73: 4004-4022.

D'Hondt S, Rutherford S, Spivack A J, 2002. Metabolic activity of subsurface life in deep-sea sediments. Science, 295: 2067-2070.

Dalal R C, 1998. Soil microbial biomass-what do the numbers really mean. Australian Journal of Experimental Agriculture, 38: 649-665.

de Wever H, Cole J R, Fettig M R, et al, 2000. Reductive dechlorination of trichloroacetic acid by Trichlorobacter thiogenes gen. nov. , sp. nov. Applied and Environmental Microbiology, 66: 2297-2301.

Dehorter B, Kontchou C Y, Blondeau R, 1992. ^{13}C NMR Spectroscopic analysis of soil humic acids recovered after incubation with some white rot fungi and actinomycetes. Soil Biology & Biochemistry, 24: 667-673.

DeLong E F, Pace N R, 2001. Environmental diversity of bacteria and archaea. Systematic Biology, 50: 470-478.

Deonarine A, Hsu-Kim H, 2009. Precipitation of mercuric sulfidenanoparticles in NOM-containing water: Implications for the natural environment. Environmental Science & Technology, 43: 2368-2373.

DiChristina T J, Fredrickson J K, Zachara J M, 2005. Enzymology of electron transport: Energy generation with geochemical consequences. In Molecular Geomicrobiology; Mineralogical Society of America: Chantilly, VA, 59: 27-52.

Dixit S, Hering J G, 2003. Comparison of arsenic (V) and arsenic (III) sorption onto iron oxide minerals: Implications for arsenicmobility. Environmental Science & Technology, 37: 4182-4189.

Dobbin P S, Butt J N, Powell A K, et al, 1999. Characterization of a flavocytochrome that is induced during the anaerobic respiration of Fe (III) by Shewanella frigidimarina NCIMB 400. Biochemical Journal, 342: 439-448.

Donald J W, Hicks M G, Richardson D J, et al, 2008. The c-type cytochrome OmcA localizes to the outer membrane upon heterologous expression in Escherichia coli. Journal of Bacteriology, 190: 5127-5131.

Doong R A, Schink B, 2002. Cysteine-mediated reductive dissolution of poorly crystalline rion (III) oxides by Geobacter sulfurreducens. Environmental Science & Technology, 36: 2939-2945.

Druschel G K, Emerson D, Sutka R, et al, 2008. Low-oxygen and chemical kinetic constraints on the geochemical niche of neutrophilic iron (II) oxidizing microorganisms. Geochimica et Cosmochimica Acta, 72: 3358-3370.

Dunnivant F M, Schwarzenbach R P, Macalady D L, 1992. Reduction of substituted nitrobenzenes in aqueous solutions containing natural organic matter. Environmental Science & Technology, 26: 2133-2141.

Elsner M, Schwarzenbach R P, Haderlein S B, 2004. Reactivity of Fe(Ⅱ)-bearing minerals toward reductive transformation of organic contaminants. Environmental Science & Technology, 38: 799-807.

Esham E C, Ye W Y, Moran M A, 2000. Identification and characterization of humic substances-degrading bacterial isolates from an estuarine environment. FEMS Microbiology Ecology, 34: 103-111.

Eusterhues K, Rumpel C, Kleber M, et al, 2003. Stabilization of soil organic matter by interactions with minerals as revealed by mineral dissolution and oxidative degradation. Organic Geochemistry, 34: 1591-1600.

Eusterhues K, Wagner F E, Hausler W, et al, 2008. Characterization of ferrihydrite-soil organic matter coprecipitates by X-ray diffraction and Mössbauer spectroscopy. Environmental Science & Technology, 42: 7891-7897.

Fendorf S E, 1995. Surface reactions of chromium in soils and waters. Geoderma, 67: 55-71.

Fiedler S, Vepraskas M J, Richardson J L, 2007. Soil redox potential: importance, field measurements, and observations. Advances in Agronomy, 94: 1-54.

Filip Z, Tesarova M, 2004. Microbial degradation and transformation of humic acids from permanent meadow and forest soils. International Biodeterioration & Biodegradation, 54: 225-231.

Fimmen R L, Cory R M, Chin Y P, et al, 2007. Probing the oxidation-reduction properties of terrestrially and microbially derived dissolved organic matter. Geochimica et Cosmochimica Acta, 71: 3003-3015

Finneran K T, Anderson R T, Nevin K P, et al, 2002. Potential for bioremediation of uranium-contaminated aquifers with microbial U(Ⅵ) reduction. Soil & Sediment Contamination, 11: 339-357.

Flynn T M, O'Loughlin E J, Mishra B, et al, 2014. Sulfur-mediated electron shuttling during bacterial iron reduction. Science, 344 (6187): 1039-1042.

Francis C A, Obraztsova A Y, Tebo B M, 2000. Dissimilatory metal reduction by the facultative anaerobe pantoea agglomerans SP1. Applied and Environmental Microbiology, 66 (2): 543-548

Fredrickson J K, Zachara J M, Kennedy D W, et al, 2004. Reduction of TcO_4^- by sediment-associated biogenic Fe(Ⅱ). Geochimica et Cosmochimica Acta, 68: 3171-3187.

Friedrich M W, Finster K W, 2014. How sulfur beats iron. Science, 344: 974-975.

Fulda B, Voegelin A, Maurer F, et al, 2013. Copper redox transformation and complexation by reduced and reoxidized soil humic acid. 1. X-ray absorption spectroscopy study. Environmental Science & Technology, 47: 10903-10911.

Gillham R W, Ohannesin S F, 1994. Enhanced degradation of halogenated aliphatics by zero-valent iron. Ground Water, 32: 958-967.

Ginder-Vogel M, Borch T, Mayes M A, et al, 2005. Chromate reduction and retention processes within arid subsurface environments. Environmental Science & Technology, 39: 7833-7839.

Ginder-Vogel M, Criddle C S, Fendorf S, 2006. Thermodynamic constraints on the oxidation of biogenic UO_2 by Fe(Ⅲ)(hydr) oxides. Environmental Science & Technology, 40: 3544-3550.

Glaus M A, Hejiman C G, Schwarzenbach R P, et al, 1992. Reduction of nitroaromatic compounds mediated by Streptomyces sp. exudates. Applied and Environmental Microbiology, 58: 1945-1951.

Gorby Y A, Lovley D R, 1992. Enzymatic uranium precipitation. Environmental Science & Technology, 26: 205-207.

Gorby Y A, Yanina S, McLean J S, et al, 2006. Electrically conductive bacterial nanowires produced by Shewanella oneidensis strain MR-1 and other microorganisms. Proceedings of the National Academy of Sciences, 103: 11358-11363.

Gorski C A, Nurmi J T, Tratnyek P G, et al, 2010. Redox behavior of magnetite: implications for contaminant reduction. Environmental Science & Technology, 44: 55-60.

Gorski C A, Scherer M M, 2009. Influence of magnetite stoichiometry on Fe(Ⅱ) uptake and nitrobenzene reduction. Environmental Science & Technology, 43: 3675-3680.

Gramss G, Ziegenhagen C, Sorge S, 1999. Degradation of soil humic extract by wood-and soil-associated fungi, bacteria, and commercial enzymes. Microbial Ecology, 37: 140-151.

Gray H B, Winkler J R, 2009. Electron flow through proteins. Chemial Physics Letters, 483: 1-9.

Grinhut T, Hadar Y, Chen Y, 2007. Degradation and transformation of humic substances by saprotrophic fungi: processes and mechanisms. Fungal Biology Reviews, 21: 179-189.

Grinhut T, Hertkorn N, Schmitt-Kopplin P, et al, 2011. Mechanisms of humic acids degradation by white rot fungi explored using ^1H NMR spectroscopy and FTICR mass spectrometry. Environmental Science & Technology, 45: 2748-2754.

Gu B, Bian Y, Miller C L, et al, 2011. Mercury reduction and complexation by natural organic matter in anoxic environments. Proceedings of the National Academy of Sciences, 25: 1479-1483.

Guggenberger G, 2005. Humification and mineralisation in soils. In: Buscot F, Varma A, eds. Microorganisms in Soils: Roles in Genesis and Functions. Springer, Berlin, Heidelberg, 85-106.

Guillaumont R, Fanghanel T, Neck V, et al, 2003. Update on the chemical thermodynamics of uranium, neptumium, plutonium, americium, and technetium. Amsterdam; Boston: Elsevier; Paris: Nuclear Energy Agency, Organisation for Economic co-Operation and Development.

Hakala J A, Chin Y P, Weber E J, 2007. Influence of dissolved organic matter and Fe(Ⅱ) on the abiotic reduction of pentachloronitrobenzene. Environmental Science & Technology, 41: 7337-7342.

Han W X, Fang J Y, Reich P B, et al, 2011. Biogeography and variability of eleven mineral elements in plant leaves across gradients of climate, soil and plant functional type in China. Ecology Letters, 14 (8), 788-796.

Harris H W, El-Naggar M Y, Bretschger O, et al, 2010. Electrokinesis is a microbial behavior that requires extracellular electron transport. Proceedings of the National Academy of Sciences of the United States of America, 107: 326-331.

Hartshorne R S, Reardon C L, Ross D, et al, 2009. Characterization of an electron conduit between bacteria and the extracellularenvironment. Proceedings of the National Academy of Sciences, 106: 22169-22174.

Hatakka A, 1994. Lignin-modifying enzymes from selected white-rot Fungi: production and role in lignin degradation. FEMS Microbiology Reviews, 13: 125-135.

Hedges J I, Blanchette R A, Weliky K, et al, 1988. Effects of fungal degra-dation on the CuO oxidation products of lignin: a controlled laboratory study. Geochimica et Cosmochimica Acta, 52: 2717-2726.

Heidelberg J F, Paulsen I T, Nelson K E, et al, 2002. Genome sequence of the dissimilatory metal ion-reducing bacterium Shewanella oneidensis. Nature Biotechnology, 20: 1118-1123.

Hernandez M E, Newman D K, 2001. Extracellular electron transfer. Cellular and Molecular Life Sciences, 58: 1562-1571.

Hernández-Montoya V, Alvarez L H, Montes-Morán M A, et al, 2012. Reduction of quinone and non-quinone redox functional groups in different humic acid samples by Geobacter sulfurreducens. Geoderma, 183-184: 25-31

Hofrichter M, Fritsche W, 1996. Depolymerization of low rank coal by extracellular fungal enzyme systems . 1. Screening for low rank-coal-depolymerizing activities. Applied Microbiology and Biotechnology, 46: 220-225.

Hohmann C, Winkler E, Morin G, et al, 2010. Anaerobic Fe (Ⅱ)-Oxidizing bacteria show as resistance and immobilize as during Fe(Ⅲ) mineral precipitation. Environmental Science & Technology, 44 (1): 94-101.

Holmes D E, Giloteaux L, Chaurasia A K, et al, 2015. Evidence of Geobacter-associated phage in auranium-contaminated aquifer. Mutidisciplinary Journal Microbial Ecology, 9: 333-346.

Holmes D E, Bond D R, O'Neil R A, et al, 2004. Microbial communities associated with electrodes harvesting electricity from a variety of aquatic sediments. Microbial Ecology, 48: 178-190.

Howarth R W, 2002. The Nitrogen Cycle. In encyclopedia of global environmental change, vol. 2, The earth system: Biological and ecological dimensions of global environmental change. Wiley: Chichester.

Hu H, Hui L, Wang Z, et al, 2013. Oxidation and methylation of dissolved elementalmercury by anaerobic bacteria. Nature Geoscience, 6: 751-754

Hug S J, Leupin O, 2003. Iron-catalyzed oxidation of arsenic (Ⅲ) by oxygen and by hydrogen peroxide: pH-dependent formation of oxidants in the Fenton reaction. Environmental Science & Technology, 37: 2734-2742.

Icopini G A, Lack J G, Hersman L E, et al, 2009. Plutonium (Ⅴ/Ⅵ) reduction by the metal-reducing bacteria Geobacter metallireducens GS-15 and Shewanella oneidensis MR-1. Applied and Environmental Microbiology, 75: 3641-3647.

Inoue K, Qian X, Moigado L, et al, 2010. Purification and characterization of OmcZ, an outer-surface, octaheme c-type cytochrome essential for optimal current production by Geobacter sulfurreducens. Applied and Environmental Microbiology, 76: 3999-4007.

IPCC, 2001. Climate Change 2001: The scientific basis. The Third Assessment Report of Working Group. Cambridge: Cambridge Univ Press.

IPCC, 2007. Climate Change 2007: The physical scientific basis. The Fourth Assessment Report of Working Group. Cambridge: Cambridge Univ Press.

IPCC, 2013. Climate Change 2013: The physical scientific basis. The Fifrth Assessment Report of Working Group. Cambridge: Cambridge Univ Press.

Ishibashi Y, Cervantes C, Silver S, 1990. Chromium reduction in Pseudomonas putida. Applied and Environmental Microbiology, 56: 2268-2270.

Jeon B H, Kelly S D, Kemner K M, et al, 2004. Microbial reduction of U(Ⅵ) at the solid water interface. Environmental Science & Technology, 38: 5649-5655.

Jeyasingh J, Philip L, 2005. Bioremediation of chromium contaminated soil: optimization of operating parameters under laboratory conditions. Journal of Hazardous Materials, 118: 113-120.

Jeyasingh J, Somasundaram V, Philip L, et al, 2010. Pilot scale studies on the remediation of chromium contaminated aquifer using bio-barrier and reactive zone technologies. Chemical Engineering Journal, 167: 206-214.

Jiang J, Bauer I, Paul A, et al, 2009. Arsenic redox changes by microbially and chemically formed semiquinone radicalsand hydroquinones in a humic substance model quinone. Environmental Science & Technology, 43: 3639-3645.

Jiang J, Kappler A, 2008. Kinetics of microbial and chemical reduction of humic substances: implications for electron shuttling. Environmental Science & Technology, 42: 3563-3569.

Jiao Y, Newman D K, 2007. The pio operon is essential for phototrophic Fe(Ⅱ) oxidation in Rhodopseudomonas palustris TIE-1. Journal of Bacteriology, 189: 1765-1773.

Jones A M, Collins R N, Rose J, et al, 2009. The effect of silica and natural organic matter on the Fe(Ⅱ)-catalysedtransformation and reactivity of Fe(Ⅲ) minerals. Geochimica et Cosmochimica Acta, 73: 4409-4422.

Jonsson S, Skyllberg U, Nilsson M B, et al, 2014. Differentiated availability of geochemical mercury pools controls methylmercury levels in estuarinesediment and biota. Nature Communications, 4624.

Kaden J, Galushko A S, Schink B, 2002. Cysteine-mediated electron transfer in syntrophic acetate oxidation by cocultures of Geobacter sulfurreducens and Wolinella succinogenes. Archives of Microbiology, 178: 53-58.

Kalbitz K, Schwesig D, Rethemeyer J, et al, 2005. Stabilization of dissolved organic matter by sorption to the mineral soil. Soil Biology & Biochemistry, 37: 1319-1331.

Kappler A, Benz M, Schink B, et al, 2004. Electron shuttlingvia humic acids in microbial iron (Ⅲ) reduction in a freshwatersediment. FEMS Microbiology Ecology, 47: 85-92.

Kappler A, Haderlein S B, 2003. Natural organic matter as reductant for chlorinated aliphatic pollutants. Environmental Science & Technology, 37: 2714-2719.

Kappler A, Straub K L, 2005. Geomicrobiological cycling of iron. Reviews in Mineralogy & Geochemistry, 59: 85-108.

Kappler A, Wuestner M L, Ruecker A, et al, 2014. Biochar as an electron shuttle between bacteria and Fe(Ⅲ) minerals. Environmental Science & Technology Letters, 1: 339-344.

Kato S, Hashimoto K, Watanabe K, 2012. Microbial interspecies electron transfer via electriccurrents through conductive minerals. Proceedings of the National Academy of Sciences, 109: 10042-10046.

Kelleher B P, Simpson A J, 2006. Humic substances in soils: Are they really chemically distinct. Environmental Science & Technology, 40: 4605-4611.

Kennedy M J, Pevear D R, Hill R J, 2002. Mineral surface control of organic carbon in black shale. Science, 295: 657-660.

Kilic N K, Stensballe A, Otzen D E, et al, 2009. Proteomic changes in response to chromium (Ⅵ) toxicity in Pseudomonas aeruginosa. Biosoure Technology, 101: 2134-2140.

Kimbrough D E, Cohen Y, Winer A M, et al, 1999. A critical assessment of chromium in the environment. Critical Reviews in Environmental Science and Technology, 29: 1-46.

King J K, Kostka J E, Frischer M E, et al, 2000. Sulfate-reducing bacteria methylate mercury at variable rates in pure culture and in marine sediments. Applied and Environmental Microbiology, 66: 2430-2437.

Kirk G, 2004. The biogeochemistry of submerged soils. John Wiley& Sons, Ltd: Chichester, 282.

Kleber M, Johnson M G, 2010. Advances in understanding the molecular structure of soil organic matter: implications for interactions in the environment. Advances in Agronomy, 106: 77-142.

Kleber M, Sollins P, Sutton R, 2007. A conceptual model of organo-mineral interactions in soils: self-assembly of organic molecular fragments into zonal structures on mineral surfaces. Biogeochemistry, 85: 9-24.

Klüpfel L, Piepenbrock A, Kappler A, et al, 2014. Humic substances as fully regenerable electron acceptors in recurrently anoxic environments. Nature Geoscience, 7: 195-200.

Kögel-Knabner I, Guggenberger G, Kleber M, et al, 2008. Organo-mineral associations in temperate soils: Integrating biology, mineralogy and organic matter chemistry. Journal of Plant Nutrition and Soil Science, 171: 61-82.

Kögel Knabner I, 2002. The macromolecular organic composition of plant and microbial residues as inputs to soil organic matter. Soil Biology & Biochemistry, 34: 139-162.

Konovalova V V, Dmytrenko G M, Nigmatullin R R, et al, 2003. Chromium (Ⅵ) reduction in a membrane bioreactor with immobilized Pseudomonas cells. Enzyme and Microbial Technology, 33: 899-907.

Kulp T R, Hoeft S E, Asao M, et al, 2008. Arsenic (Ⅲ) fuels anoxygenic photosynthesis in hot spring biofilms from Mono Lake. California. Science, 321: 967-970.

Kunapuli U, Jahn M K, Lueders T, et al, 2010. Desulfitobacterium aromaticivorans sp. nov. and Geobacter toluenoxydans sp. nov., ironreducing bacteria capable of anaerobic degradation of monoaromatic hydrocarbons. International Journal of Systematic and Evolutionary Microbiology, 60: 686-695.

Leang C, Qian X, Mester T, et al, 2010. Alignment of the c-type cytochrome OmcS along pili of Geobacter sulfurreducens. Applied and Environmental Microbiology, 76: 4080-4084.

Lebedeva Lialikova, 1979. Crocite reduction by a culture of Pseudomonas chromatophila sp. nov. Mikrobiologiia, 48: 517-522.

Lehmann J, Solomon D, Kinyangi J, et al, 2008. Spatial complexity of soil organic matter forms at nanometer scales. Nature Geoscience, 1: 238-242.

Leys D, Meyer T E, Tsapin A S, et al, 2002. Crystal structures at atomic resolution reveal the novel concept of "electron-harvesting" as a role for the small tetraheme cytochrome c. Journal of Biological Chemistry, 277: 35703-35711.

Lies D P, Hernandez M E, Kappler A, et al, 2005. Shewanella oneidensis MR-1 uses overlapping pathways for iron reduction at a distance and by direct contact under conditions relevant for biofilms. Applied and Environmental Microbiology, 71: 4414-4426.

Liger E, Charlet L, Van Cappellen P, 1999. Surface catalysis ofuranium (Ⅵ) reduction by iron (Ⅱ). Geochimica et Cosmochimica Acta, 63: 2939-2955.

Lin K, Liu W, Gan J, 2009. Oxidative removal of bisphenol a by manganese dioxide: efficacy, products, and pathways. Environmental Science & Technology, 43: 3860-3864.

Liu F, Rotaru A E, Shrestha P M, et al, 2012. Promoting direct interspecies electron transfer with activated carbon. Energy & Environmental Science, 5: 8982-8989.

Lloyd J R, Chesnes J, Glasauer S, et al, 2002. Reduction of actinides and fission products by Fe (Ⅲ)-reducing bacteria. Geomicrobiology Journal, 19: 103-120.

Lloyd J R, Lovley D R, Macaskie L E, et al, 2003. Biotechnological application of metal-reducing microorganisms. Advances in Applied Microbiology, 53: 85-128.

Lloyd J R, Mabbett A N, Williams D R, et al, 2001. Metal reduction by sulphate-reducing bacteria: physiological diversity and metal specificity. Hydrometallurgy, 59: 327-337.

Lloyd J R, Sole V A, Van Praagh C V, et al, 2000. Direct and Fe (II)-mediated reduction of technetium by Fe (III)-reducing bacteria. Applied and Environmental Microbiology, 66: 3743-3749.

Lohmayer R, Kappler A, Loesekann-Behrens T, et al, 2014. Sulfur species as redox partners and electron shuttles for ferrihydrite reduction by Sulfurospirillum deleyianum. Applied and Environmental Microbiology, 80: 3141-3149.

Lovley D R, Baedecker M J, Lonergan D J, et al, 1989. Oxidation of aromatic contaminants coupled to microbial iron reduction. Nature, 339: 297-300.

Lovley D R, Coates J D, Blunt-Harris E L, et al, 1996. Humic substances as electron acceptors for microbial respiration. Nature, 382: 445-448.

Lovley D R, Holmes D E, Nevin K P, et al, 2004. Dissimilatory Fe(III) and Mn(IV) reduction. Advances in Microbial Physiology, 49: 219-286.

Lovley D R, Lonergan D J, 1990. Anaerobic oxidation of toluene, phenol, and p-cresol by the dissimilatory iron-reducing organism, GS-15. Applied and Environmental Microbiology, 56: 1858-1864.

Lovley D R, Phillips E J, 1992. Reduction of uranium by Desulfovibrio desulfuricans. Applied and Environmental Microbiology, 58: 850-856.

Lovley D R, Phillips E J P, Gorby Y A, et al, 1991. Microbial reduction of uranium. Nature, 350: 413-416.

Lovley D R, Phillips E J P, 1994. Reduction of chromate by Desulfovibrio vulgaris and its c3 cytochrome. Applied and Environmental Microbiology, 60: 726-728.

Lovley D R, Phillips E J P, 1988. Novel mode of microbial energy metabolism: organic carbon oxidation coupled to dissimilatory reduction of iron or manganese. Applied and Environmental Microbiology, 54: 1472-1480.

Lovley D R, Stolz J F, Nord G L J, et al, 1987. Anaerobic production of magnetite by a dissimilatory iron-reducing microorganism. Nature, 330: 252-254.

Lovley D R, Widman P K, Woodward J C, et al, 1993. Reduction of uranium by cytochrome c3 of Desulfovibrio vulgaris. Applied and Environmental Microbiology, 59: 3572-3576.

Lovley D R, Woodward J C, Chapelle F H, 1994. Stimulated anoxic biodegradation of aromatic hydrocarbons using Fe(III) ligands. Nature, 370: 128-131.

Lovley D R, 1993. Dissimilatory metal reduction. Annual Review of Microbiology, 47: 263-290.

Lovley D R, Kashefi K, Vargas M, et al, 2000. Reduction of humic substances and Fe(III) by hyperthermophilic microorganisms. Chemical Geology, 169: 289-298.

Lovley D R, Ueki T, Zhang T, et al, 2011. Geobacter: the microbe electric's physiology, ecology and practical applications. Advances in Microbial Physiology, 59: 1-100.

Lucassen E, Smolders A J P, Van der Salm A L, et al, 2004. High groundwater nitrate concentrations inhibiteutrophication of sulphate-rich freshwater wetlands. Biogeochemistry, 67: 249-267.

Luther III G W, Rickard D T, 2005. Metal sulfide cluster complexesand their biogeochemical importance in the environment. Journal of Nanoparticle Research, 7: 389-407.

Lyngkilde J, Christensen T H, 1992. Fate of organic contaminants in the redox zones of a landfill leachate pollution plume (Vejen, Denmark). Journal of Contaminant Hydrology, 10: 291-307.

Magnuson T, Swenson M, Paszczynski A, et al, 2010. Proteogenomic and functional analysis of chromate reduction in Acidiphilium cryptum JF-5, an Fe (Ⅲ)-respiring acidophile. Biometals, 23: 1129-1138.

Malvankar N S, Vargas M, Nevin K P, et al, 2011. Tunable metallic-like conductivity in microbialnanowire networks. Nature, 6: 573-579.

Malvankar N S, Lovley D R, 2012. Microbial nanowires: A new paradigm for biological electron transfer and bioelectronics. Chem Sus Chem, 5: 1039-1046.

Malvankar N S, Lovley D R, 2014. Microbial nanowires for bioenergy applications. Current Opinion in Biotechnology, 27: 88-95.

Marsili E, Baron D B, Shikhare I D, et al, 2008. Shewanella secretes flavins that mediate extracellular electron transfer. Proceedings of the National Academy of Sciences, 105: 3968-3973.

Martinez C M, Alvarez L H, Celis L B, et al, 2013. Humus-reducing microorganisms and their valuable contribution in environmental processes. Applied Microbiology and Biotechnology, 97: 10293-10308.

Maurer F, Christl I, Fulda B, et al, 2013. Copper redox transformation and complexation by reduced and oxidized soil humic acid. 2. Potentiometric titrations and dialysis cell experiments. Environmental Science & Technology, 47: 10912-10921.

Maurer F, Christl I, Hoffmann M, et al, 2012. Reduction and reoxidation of humic acid: Influence on speciation of cadmium and silver. Environmental Science & Technology, 46: 8808-8816.

Mayer L M, Schick L L, Hardy K R, et al, 2004. Organic matter in small mesopores in sediments and soils. Geochimica et Cosmochimica Acta, 68: 3868-3872.

McCarthy J F, Ilavsky J, Jastrow J D, et al, 2008. Protection of organic carbon in soil microaggregates occur via restructuring of aggregate porosity and filling pores with accumulating organic matter. Geochimica et Cosmochimica Acta, 72: 4725-4744.

McCormick M L, Adriaens P, 2004. Carbon tetrachloride transformation on the surface of nanoscale biogenic magnetite particles. Environmental Science & Technology, 38: 1045-1053.

McKeague J A, Cheshire M V, Andreux F, et al, 1986. Organo-mineral complexes in relation to pedogenesis. In: Huang P M, Schnitzer M, eds. Interactions of soil minerals with natural organics and microbes. Soil Science Society of America, Madison, WI. 549-592.

McLean J S, Pinchuk G E, Geydebrekht O V, et al, 2008. Oxygen-dependent autoaggregation in Shewanella oneidensis MR-1. Environmental Microbiology, 10: 1861-1876.

Methé B A, Nelson K E, Eisen J A, et al, 2003. Genome of Geobacter sulfurreducens: metal reduction in subsurface environments. Science, 302: 1967-1969.

Mikutta R, Kleber M, Torn M S, et al, 2006. Stabilization of soil organic matter: Association with minerals or chemical recalcitrance? Biogeochemistry, 77: 25-56.

Mikutta R, Mikutta C, Kalbitz K, et al, 2007. Biodegradation of forest floor organic matter bound to minerals via different binding mechanisms. Geochimica et Cosmochimica Acta, 71: 2569-2590.

Miyata N, Tani Y, Maruo K, et al, 2006. Manganese (Ⅳ) oxide production by Acremonium sp strain KR21-2 and extracellular Mn (Ⅱ) oxidase activity. Applied and Environmental Microbiology, 72: 6467-6473.

Moore C M, 2014. Microbial proteins and oceanic nutrient cycles. Science, 345: 1120-1121.

Morel F O M M, Kraepiel A M L, Amyot M, 1998. The chemical cycle and bioaccumulation of mercury. Annual Review of Ecology and Systematics, 29: 543-566.

Morse J W, Luther G W, 1999. Chemical influences on tracemetal-sulfide interactions in anoxic sediments. Geochimica et Cosmochimica Acta, 63: 3373-3378.

Myers C R, Nealson K H, 1988. Microbial reduction of manganese oxides: interactions with iron and sulfur. Geochimica et Cosmochimica Acta, 52: 2727-2732.

Myers C R, Nealson K H, 1990. Respiration-linked proton translocation coupled to anaerobic reduction of manganese (Ⅳ) and iron (Ⅲ) in Shewanella putrefaciens MR-1. Journal of Bacteriology, 172: 6232-6238.

Myers J M, Myers C R, 2002. MtrB is required for proper incorporation of the cytochromes omcA and omcB into the outer membrane of Shewanella putrefaciens MR-1. Applied and Environmental Microbiology, 68: 5585-5594.

Nagarajan H, Embree M, Rotaru A E, et al, 2013. Characterization and modelling of interspecies electron transfer mechanisms and microbialcommunity dynamics of a syntrophic association. Nature Communications, 2809.

Nakamura K, Hagimine M, Sakai M, et al, 1999. Removal of mercury from mercurycontaminated sediments using a combined method of chemical leaching and volatilization of mercury by bacteria. Biodegradation, 10: 443-447.

Nancharaiah Y V, Dodge C, Venugopalan V P, et al, 2010. Immobilization of Cr(Ⅵ) and its reduction to Cr (Ⅲ) phosphate by granular biofilms comprising a mixture of microbes. Applied and Environmental Microbiology, 76: 2433-2438.

Nevin K P, Lovley D R, 2002. Mechanisms for accessing insoluble Fe(Ⅲ) oxide during dissimilatory Fe(Ⅲ) reduction by Geothrix fermentans. Applied and Environmental Microbiology,

68: 2294-2299.

Newman D K, Kolter R, 2000. A role for excreted quinones in extracellular electron transfer. Nature, 405: 94-97.

Newsome L, Morris K, Lloyd J R, 2014. The biogeochemistry and bioremediation of uranium and other priority radionuclides. Chemical Geology, 363: 164-184.

Newton G J, Mori S, Nakamura R, et al, 2009. Analyses of current-generating mechanisms of Shewanella loihica PV-4 and Shewanella oneidensis MR-1 in microbial fuel cells. Applied and Environmental Microbiology, 75 (24): 7674-7681.

Nielsen L P, Risgaard-Petersen N, Fossing H, et al, 2010. Electric currents couple spatially separated biogeochemical processes in marine sediment. Nature, 463: 1071-1074.

Nielsen P H, Albrechtsen H J R, Heron G, et al, 1995. In situ and laboratory studies on the fate of specific organic compounds in an anaerobic landfill leachate plume, 1. Experimental conditions and fate of phenolic compounds. Journal of Contaminant Hydrology, 20: 27-50.

Nierop K G J, Jansen B, Verstraten J A, 2002. Dissolved organic matter, aluminium and iron interactions: precipitation induced by metal/carbon ratio, pH and competition. Science of the Total Environment, 300: 201-211.

O'Hannesin S F, Gillham R W, 1998. Long-term performance of an in situ "iron wall" for remediation of VOCs. Ground Water, 36: 164-170.

O'Loughlin E J, Kelly S D, Kemner K M, et al, 2003. Reduction of Ag (Ⅰ), Au (Ⅲ), Cu (Ⅱ), and Hg (Ⅱ) by Fe (Ⅱ) /Fe (Ⅲ) hydroxysulfate green rust. Chemosphere, 53: 437-446.

Oades J M, 1993. The role of biology in the formation, stabilization and degradation of soil structure. Geoderma, 56: 377-400.

Obuekwe C O, Westlake W, Cook F, 1981. Effect of nitrate on reduction of ferric iron by a bacterium isolated from crude oil. Canadian Journal of Microbiology, 27: 692-697.

Okamoto A, Hashimoto K, Nealson K H, et al, 2013. Rate enhancement of bacterial extracellular electron transport involves bound flavin semiquinones. Proceedings of the National Academy of Sciences, 110: 7856-7861.

Ortiz-Bernad I, Anderson R T, Vrionis H A, et al, 2004. Resistance of solid-phase U(Ⅵ) to microbial reduction during in situ bioremediation of uranium-contaminated groundwater. Applied and Environmental Microbiology, 70: 7558-7560.

Paul S, 2002. Humus: Still a Mystery. Northeast Organic Farming Association.

Peth S, Horn R, Beckmann F, et al, 2008. Three-dimensional quantification of intra-aggregate pore-space features using synchrotron-radiation-based microtomography. Soil Science Society of America Journal, 72: 897-907.

Piccolo A, 2001. The supramolecular structure of humic substances. Soil Science, 166: 810-832.

Piepenbrock A, Schroöder C, Kappler A, 2014. Electron transfer from humic substances to biogenic and abiogenic Fe(Ⅲ) oxyhydroxide minerals. Environmental Science & Technology, 48: 1656-1664.

Pirbadian S, Barchinger S E, Leung K M, et al, 2014. Shewanella oneidensis MR-1 nanowires are outer membrane and periplasmic extensions of the extracellular electron transport components. Proceedings of the National Academy of Sciences, 111 (35): 12883-12888.

Pitts K E, Dobbin P S, Reyes-amirez F, et al, 2003. Characterization of the Shewanella oneidensis MR-1 decaheme cytochrome mtrA. Journal of Biological Chemistry, 278: 27758-27765.

Polizzotto M L, Kocar B D, Benner S G, et al, 2008. Near-surface wetland sediments as a source of arsenic release to ground water in Asia. Nature, 454: 505-508.

Post J E, 1999. Manganese oxide minerals: Crystal structures andeconomic and environmental significance. Proceedings of the National Academy of Sciences, 96: 3447-3454.

Powell B A, Duff M C, Kaplan D I, et al, 2006. Plutoniumoxidation and subsequent reduction by Mn(Ⅳ) minerals in Yucca Mountain tuff. Environmental Science & Technology, 40: 3508-3514.

Qi B C, Aldrich C, Lorenzen L, et al, 2004. Degradation of humic acids in a microbial film consortium from landfill compost. Industrial & Engineering Chemistry Research, 43: 6309-6316.

Qian X, Mester T, Morgado L, et al, 2010. Biochemical characterization of purified OmcS, a c-type cytochrome required for insoluble Fe(Ⅲ) reduction in Geobacter sulfurreducens. Biochimica et Biophysica Acta, 1807: 404-412.

Rai D, Eary L E, Zachara J M, 1989. Environmental chemistry of chromium. Science of the Total Environment, 86: 15-23.

Ramana C V, Sasikala C, 2009. Albidoferax, a new genus of Comamonadaceae and reclassification of Rhodoferax ferrireducens (Finneran et al. 2003) as Albidoferax ferrireducens comb. The Journal of general and applied microbiology, 55: 301-304.

Ratasuk N, Nanny M A, 2007. Characterization and quantification of reversible redox sites in humic substances. Environmental Science & Technology, 41 (22): 7844-7850

Reardon C L, Dohanlkova A C, Nachimuthu P, et al, 2010. Role of outer-membrane cytochromes MtrC and OmcA in the biomineralization of ferrihydrite by Shewanella oneidensis MR-1. Geobiology, 8: 6-68.

Redman A D, Macalady D L, Ahmann D, 2002. Natural organicmatter affects arsenic speciation and sorption onto hematite. Environmental Science & Technology, 36: 2889-2896.

Reguera G, McCarthy K D, Mehta T, et al, 2005. Extracellular electron transfer via microbial nanowires. Nature, 435: 1098-1101.

Reyes-Ramirez F, Dobbin P, Sawers G, et al, 2003. Characterization of transcriptional regulation of Shewanella frigidimarina Fe (Ⅲ)-induced flavocytochrome c reveals a novel iron-responsive gene regulation system. Journal of Bacteriology, 185: 4561-4571.

Rezacova V, Hrselova H, Gryndlerova H, et al, 2006. Modifications of degradation-resistant soil organic matter by soil saprobic microfungi. Soil Biology & Biochemistry, 38: 2292-2299.

Richardson D J, Butt J N, Fredrickson J K, et al, 2012. The "porin-cytochrome" model for microbe-to-mineral electron transfer. Molecular Microbiology, 85: 201-212.

Roden E, Kappler A, Bauer I, et al, 2010. Extracellular electron transfer through microbial reduction of solid-phase humic substances. Nature Geoscience, 3: 417-421.

Roh Y, Chon C-M, Moon J-W, 2007. Metal reduction and biomineralization by an alkaliphilic metal-reducing bacterium, Alkaliphilus metalliredigens (QYMF). Geosciences Journal, 11: 415-423.

Roling W F M, van Breukelen B M, Braster M, et al, 2001. Relationships between microbial community structure and hydrochemistry in a landfill leachate-polluted aquifer. Applied and Environmental Microbiology, 67: 4619-4629.

Romanenko V I, Koren'kov V N, 1977. Pure culture of bacteria using chromates and bichromates as hydrogen acceptors during development under anaerobic conditions. Mikrobiologiia, 46: 414-417.

Rooney-Varga J N, Anderson R T, Fraga J L, et al, 1999. Microbial communities associated with anaerobic benzene degradation in a petroleum-contaminated aquifer. Applied and Environmental Microbiology, 65: 3056-3063.

Ross D E, Ruebush S S, Brantley S L, et al, 2007. Characterization of protein-protein interactions involved in iron reduction by Shewanella oneidensis MR-1. Applied and Environmental Microbiology, 73: 5797-5808.

Rosenbaum M A, Bar H Y, Beg Q, et al, 2012. Transcriptional analysis of shewanella oneidensis MR-1 with an electrode compared to Fe(Ⅲ) citrate or oxygen as terminal electron acceptor. Plos ONE, 7: e30827.

Rosso K M, Zachara J M, Fredrickson J K, et al, 2003. Nonlocal bacterial electron transfer to hematite surfaces. Geochimica et Cosmochimica Acta, 67: 1081-1087.

Rotarua A E, Shresthaa P M, Liu F, et al, 2014. Direct interspecies electron transfer between Geobacter metallireducens and methanosarcina barkeri. Applied and Environmental Microbiology, 80 (15), 4599-4605.

Saiz-Jimenez C, 1996. The chemical structure of humic substances: recent advances. In: Piccolo A, ed. Humic Substances in Terrestrial Ecosystems. Elsevier, Amsterdam, 1-44.

Saouter E, Gillman M, Barkay T, 1995. An evaluation of mer-specified reduction of ionic mercury as a remedial tool of a mercury-contaminated freshwater pond. Journal of Industrial Microbiology & Biotechnology, 14: 343-348.

Schaefer J K, Morel F M M, 2009. High methylation rates of mercury bound tocysteine by Geobacter sulfurreducens. Nature geoscience, 2: 123-126.

Schmidt M W I, Torn M S, Abiven S, et al, 2011. Persistence of soil organic matter as an ecosystem property. Nature, 478: 49-56.

Schnitzer M, Ripmeester J A, Kodama H, 1988. Characterization of the organic-matter associated with a soil clay. Soil Science, 145: 448-454.

Schröder U, 2007. Anodic electron transfer mechanisms in microbial fuel cells and their energy efficiency. Physical Chemistry Chemical Physics, 9: 2619-2629.

Schwesig D, Kalbitz K, Matzner E, 2003. Effects of aluminium on the mineralization of dissolved organic carbon derived from forest floors. European Journal of Soil Science, 54: 311-322.

Scott D T, McKnight D M, Blunt Harris E L, et al, 1998. Quinone moieties act as electron acceptors in the reduction of humic substances by humics-reducing microorganisms. Environmental Science & Technology, 32: 2984-2989.

Selesi D, Pattis I, Schmid M, et al, 2007. Quantification of bacterial Rubis CO genes in soil by cbbL targeted real-time PCR. Journal of Microbiological Methods, 69: 497-503.

Senko J M, Mohamed Y, Dewers T A, et al, 2005. Role for Fe(III) minerals in nitrate-dependent microbial U(IV) oxidation. Environmental Science & Technology, 39: 2529-2536.

Shaik S, 2010. Biomimetic chemistry: Iron opens up to high activity. Nature Chemistry, 2: 347-349.

Shelobolina E, Coppi M, Korenevsky A, et al, 2007. Importance of c-type cytochromes for U(VI) reductionby Geobacter sulfurreducens. BMC Microbiology, 7: 16.

Shi L, Chen B W, Wang Z M, et al, 2006. Isolation of a high-affinity functional protein complex between OmcA and MtrC: Two outer m cytochromes of Shewanella oneidensis MR-1. Journal of Bacteriology, 188 (13): 4705-4714.

Shi L, Richardson D J, Wang Z, et al, 2009. The roles of outer membrane cytochromes of Shewanella and Geobacter in extracellular electron transfer. Environmental Microbiology Reports, 1: 220-227.

Shi L, Rosso K M, Clarke T A, et al, 2012. Molecular underpinnings of Fe(III) oxide reduction by Shewanella oneidensis MR-1. Frontiers in Microbiology, 3: 50.

Shi L, Squier T C, Zachara J M, et al, 2007. Respiration of metal (hydr) oxides by Shewanella and Geobacter: A key role for multihaem c-type cytochromes. Molecular Microbiology, 65: 12-20.

Simpson A J, Song G X, Smith E, et al, 2007. Unraveling the structural components of soil humin by use of solution-state nuclear magnetic resonance spectroscopy. Environmental Science & Technology, 41: 876-883.

Six J, Bossuyt H, Degryze S, et al, 2004. A history of research on the link between (micro) aggregates, oil biota, and soil organic matter dynamics. Soil & Tillage Research, 79: 7-31.

Six J, Conant R T, Paul E A, et al, 2002. Stabilization mechanisms of soil organic matter: Implications for C-saturation of soils. Plant Soil, 241: 155-176.

Six J, Elliott E T, Paustian K, 2000. Soil macroaggregate turnover and micro-aggregate formation: A mechanism for C sequestration under no-tillage agriculture. Soil Biology & Biochemistry, 32: 2099-2103.

Smith W L, 2001. Hexavalent chromium reduction and precipitation by sulphate-reducing bacterial biofilms. Environmental Geochemistry and Health, 23: 297-300.

Snoeyenbos-West O L, Nevin K P, Anderson R T, et al, 2000. Enrichment of Geobacter species in response to stimulation of Fe(Ⅲ) reduction in sandy aquifer sediments. Microbial Ecology, 39: 153-167.

Sollins P, Homann P, Caldwell B A, 1996. Stabilization and destabilization of soil organic matter: Mechanisms and controls. Geoderma, 74: 65-105.

Sposito G, Struyk Z, 2001. Redox properties of standard humic acids. Geoderma, 102: 329-346.

Sposito G, 2011. Electron shuttling by natural organic matter: Twenty years after. In: Tratnyek P G, Grundl T J, Haderlein SB (eds) Aquatic redox chemistry. American Chemical Society, Washington, DC, 113-127.

Staats M, Braster M, Röling W F M, 2011. Molecular diversity and distribution of aromatic hydrocarbon-degrading anaerobes across a landfill leachate plume. Environmental Microbiology, 13: 1216-1227.

Stevenson F J, 1994. Humus Chemistry. New York: John Wiley & Sons.

Stevenson F J, 1982. Humus chemistry: Genesis, composition, reactions. New York: John Wiley & Sons.

Stewart B D, Nico P S, Fendorf S, 2009. Stability of uranium incorporated into Fe (Hydr) oxides under fluctuating redox conditions. Environmental Science & Technology, 43: 4922-4927.

Sutton R, Sposito G, 2005. Molecular structure in soil humic substances: The new view. Environmental Science & Technology, 39: 9009-9015.

Tan S L J, Webster R D, 2012. Electrochemically induced chemically reversible proton-coupled electron transfer reactions of riboflavin (vitamin B2). Journal of the American Chemical Society, 134: 5954-5964.

Tebo B M, Bargar J R, Clement B G, et al, 2004. Biogenic manganeseoxides: Properties and mechanisms of formation. Annual Review of Earth and Planetary Sciences, 32: 287-328.

Tebo B M, Obraztsova A Y, 1998. Sulfate-reducing bacterium grows with Cr(Ⅵ), U(Ⅵ), Mn(Ⅳ), and Fe(Ⅲ) as electron acceptors. FEMS Microbiology Letters, 162: 193-198.

Thamdrup B, Fossing H, Jørgensen B B, 1994. Manganese, iron and sulfur cycling in a coastal marine sediment, Aarhus Bay, Denmark. Geochimica et Cosmochimica Acta, 58: 5115-5129.

Tobler N B, Hofstetter T B, Straub K L, et al, 2007. Iron-mediated microbial oxidation and abiotic reduction of organic contaminants under anoxic conditions. Environmental Science & Technology, 41: 7765-7772.

Tratnyek P G, Macalady D L, 1989. Abiotic reduction of nitro aromatic pesticides in anaerobic laboratory systems. Jounal of Agricultural and Food Chemistry, 37: 248-254.

Tripathi A G A, 2002. Bioremediation of toxic chromium from electroplating effluent by chromate-reducing Pseudomonas aeruginosa A2Chr in two bioreactors. Applied Microbiology and Biotechnology, 58: 416-420.

Trompowsky P M, de Melo Benites V, Madari B E, et al, 2005. Characterization of humic like substances obtained by chemical oxidation of eucalyptus charcoal. Organic Geochemistry, 36: 1480-1489.

Tufano K J, Reyes C, Saltikov C W, et al, 2008. Reductive processes controlling arsenic retention: Revealing the relative importance of iron and arsenic reduction. Environmental Science & Technology, 42: 8283-8289.

Tunega D, Gerzabek M H, Haberhauer G, et al, 2007. Formation of 2,4-D complexes on montmorillonites an ab initio molecular study. European Journal of Soil Science, 58: 680-691.

Turick C E, Graves C, Apel W A, 1998. Bioremediation potential of Cr (Ⅵ)-contaminated soil using indigenous microorganisms. Bioremediation Journal, 2: 1-6.

Turick D E, Tisa L S, Caccavo F, 2002. Melanin production and use as a soluble electron shuttle for Fe(Ⅲ) oxide reduction and as a terminal electron acceptor by Shewanella algae BrY. Applied and Environmental Microbiology, 68: 2436-2444.

Van der Zee F P, Cervantes F J, 2009. Impact and application of electron shuttles on the redox (bio) transformation of contaminants: A review. Biotechnology Advances, 27: 256-277.

Van Nooten T, Springael D, Bastiaens L, 2008. Positive impact of microorganisms on the performance of laboratory-scale permeable reactive iron barriers. Environmental Science & Technology, 42: 1680-1686.

Vanengelen M, Peyton B, Mormile M, et al, 2008. Fe(Ⅲ), Cr(Ⅵ), and Fe(Ⅲ) mediated Cr(Ⅵ) reduction in alkaline media using a Halomonas isolate from Soap Lake, Washington. Biodegradation, 19: 841-850.

Vikesland P J, Heathcock A M, Rebodos R L, et al, 2007. Particle size and aggregation effects on magnetite reactivity toward carbon tetrachloride. Environmental Science & Technology, 41: 5277-5283.

Vikesland P J, Valentine R L, 2002. Iron oxide surface-catalyzed oxidation of ferrous iron by monochloramine: Implications of oxide type and carbonate on reactivity. Environmental Science & Technology, 36: 512-519.

von Canstein H, Li Y, Leonhauser J, et al, 2002. Spatially oscillating activity and microbial succession of mercury-reducing biofilms in a technical-scale bioremediation system. Applied and Environmental Microbiology, 68: 1938-1946.

von Canstein H, Ogawa J, Shimizu S, et al, 2008. Secretion of flavins by Shewanella species and their role in extracellular electron transfer. Applied and Environmental Microbiology, 74: 615-

623.

von Lützow M, Kögel-Knabner I, Ekschmitt K, et al, 2006. Mechanisms for organic matter stabilization in temperate soils-a synthesis. European Journal of Soil Science, 57: 426-445.

Vondrasek J, Bendova L, Kliisak V, et al, 2005. Unexpectedly strong energy stabilization inside the hydrophobic core of small protein rubredoxin mediated by aromatic residues: correlated ab initio quantum chemical calculations. Journal of the American Chemical Society, 127: 2615-2619.

Vrionis H A, Anderson R T, Ortiz-Bernad I, et al, 2005. Microbiological and geochemical heterogeneity in anin situ uranium bioremediation field site. Applied and Environmental Microbiology, 71: 6308-6318.

Wagai R, Mayer L M, Kitayama K, 2009. Extent and nature of organic coverage of soil mineral surfaces assessed by gas sorption approach. Geoderma, 149: 152-160.

Wagner-Döbler I, von Canstein H, Li Y, et al, 2000. Removal of mercury from chemical wastewater by microoganisms in technical scale. Environmental Science & Technology, 34: 4628-4634.

Wagner-Döbler I, 2003. Pilot plant for bioremediation of mercury-containing industrial wastewater. Applied Microbiology and Biotechnology, 62: 124-133.

Waldemer R H, Tratnyek P G, 2005. Kinetics of contaminant degradation by permanganate. Environmental Science & Technology, 40, 1055-1061.

Wang G, Li J, Liu X, et al, 2013. Variations in carbon isotope ratios of plants across a temperature gradient along the 400 mm isoline of mean annual precipitation in north China and their relevance to paleovegetation reconstruction. Quaternary Science Reviews, 63: 83-90.

Wang Y T, 2000. Microbial reduction of Cr(Ⅵ). In: Loveley DR (ed.) Environmental microbemetal interactions, 225-235.

Wani R, Kodam K, Gawai K, et al, 2007. Chromate reduction by Burkholderia cepacia MCMB-821, isolated from the pristine habitat of alkaline crater lake. Applied Microbiology and Biotechnology, 75: 627-632.

Watanabe K, Manefield M, Lee M, et al, 2009. Electron shuttles in biotechnology. Current Opinion in Biotechnology, 20: 633-641.

Weber F A, Voegelin A, Kaegi R, et al, 2009a. Contaminant mobilization by metallic copper and metal sulphide colloids in flooded soil. Nature Geoscience, 2: 267-271.

Weber F A, Voegelin A, Kretzschmar R, 2009b. Multi-metal contaminantdynamics in temporarily flooded soil under sulfatelimitation. Geochimica et Cosmochimica Acta, 73: 5513-5527.

Weber K A, Achenbach L A, Coates J D, 2006. Microorganisms pumping iron: anaerobic microbial iron oxidationand reduction. Nature, 4: 752-764.

White G F, Shi Z, Shi L, et al, 2013. Rapid electron exchange between surface-exposed bacterial cytochromes and Fe(Ⅲ) minerals. Proceedings of the National Academy of Sciences, 110:

6346-6351.

Wiesenberg G L B, Schwarzbauer J, Schwark L, et al, 2008. Plant and soil lipid modifications under elevated atmospheric CO_2 conditions Ⅱ: Stable carbon isotopic values ($\delta^{13}C$) and turnover. Organic Geochemistry, 39: 103-117.

Williams A G B, Gregory K B, Parkin G F, et al, 2005. Hexahydro-1,3,5-trinitro-1,3,5-triazine transformation by biologically reduced ferrihydrite: evolution of Fe mineralogy, surface area, and reaction rates. Environmental Science & Technology, 39: 5183-5189.

Williams A G B, Scherer M M, 2004. Spectroscopic evidence for Fe(Ⅱ)-Fe(Ⅲ) electron transfer at the iron oxide-water interface. Environmental Science & Technology, 38: 4782-4790.

Williams K H, Bargar J R, Lloyd J R, et al, 2013. Bioremediation of uranium-contaminated groundwater: A systems approach to subsurface biogeochemistry. Current Opinion in Biotechnology, 24: 489-497.

Williams K H, Long P E, Davis J A, et al, 2011. Acetate availabilityand its influence on sustainable bioremediation of uranium-contaminated groundwater. Geomicrobiology Journal, 28: 519-539.

Willmann G, Fakoussa R M, 1997. Biological bleaching of watersoluble coal macromolecules by a basidiomycete strain. Applied Microbiology and Biotechnology, 47: 95-101.

Wischgoll S, Heintz D, Peters F, et al, 2005. Gene clusters involved in anaerobic benzoate degradation of Geobacter metallireducens. Molecular Microbiology, 58: 1238-1252.

Wu W M, Carley J, Fienen M, et al, 2006. Pilot-scale in situ bioremediation of uranium in a highly contaminated aquifer. 1. Conditioning of a treatment zone. Environmental Science & Technology, 40: 3978-3985.

Wu W M, Carley J, Luo J, et al, 2007. In situ bioreduction of uranium (Ⅵ) tosubmicromolar levels and reoxidation by dissolved oxygen. Environmental Science & Technology, 41: 5716-5723.

Yao W, Millero F J, 1996. Oxidation of hydrogen sulfide by hydrous Fe(Ⅲ) oxides in seawater. Marine Chemistry, 52: 1-16.

Zeng X C, Hu H, Hu X Q, et al, 2009. Calculating solution redox free energies with ab initio quantum mechanical/molecular mechanical minimum free energy path method. Journal of Chemical Physic, 130: 164111.

Zhang C, Katayama A, 2012. Humin as an electron mediator for microbial reductive dehalogenation. Environmental Science & Technology, 46 (12): 6575-6583.

Zhang H, Chen W R, Huang C H, 2008. Kinetic modeling of oxidation of antibacterial agents by manganese oxide. Environmental Science & Technology, 42: 5548-5554.

Zhang T, Tremblay P L, Chaurasia A K, et al, 2013. Anaerobic benzene oxidation via phenol in Geobacter metallireducens. Appl Environmental Microbiology, 79: 7800-7806.

Zheng W, Lin H, Mann B F, et al, 2013. Oxidation of dissolved elemental mercury by thiol compounds under anoxic conditions. Environmental Science & Technology，47：12827-12834.

窦森，2010. 土壤有机质. 北京：科学出版社.

科诺诺娃，1959. 土壤有机质：有机质的本性及其在土壤形成过程和土壤肥力上的作用. 陈思健，译. 北京：科学出版社.

李阜棣，2003. 微生物学. 北京：中国农业出版社.

熊田恭一，1981. 土壤有机质的化学. 李庆荣，译. 北京：科学出版社.

第2章 土壤原位固相腐殖质电子转移能力

腐殖质是土壤中天然有机质的重要组成部分,在缺氧条件下,腐殖质可以接受来自各种不同微生物的电子,例如铁还原菌(Lovley et al., 1996)、硫酸盐还原菌(Cervantes et al., 2002)和酵母菌(Benz et al., 1998)。一旦还原,腐殖质随后可以作为在土壤和沉积物中不易被微生物接近的 Fe(Ⅲ) 氧化物的电子供体而经历不同的氧化还原过程,从而介导它们的异化还原过程(Lovley et al., 1996; Wolf et al., 2009),也可作为各种氧化还原活性有机和无机污染物 [例如氯化化合物(Borch et al., 2010; Kappler and Haderlein, 2003)、硝基苯(Dunnivant et al., 1992; Van der Zee and Cervantes, 2009)、U(Ⅵ)(Gu and Chen, 2003) 和 Cr(Ⅵ)(Nakayasu et al., 1999; Wittbrodt and Palmer, 1997)] 的电子供体来影响其氧化还原转化和形成。

腐殖质在微生物和末端电子受体之间能够产生电子穿梭作用是因为其结构中存在各种氧化还原活性官能团(Scott et al., 1998; Struyk and Sposito, 2001)。电子自旋共振测量提供了最为直接的证据,表明腐殖质中的醌基团是微生物还原过程中主要的电子受体官能团(Scott et al., 1998)。基因证据表明,在腐殖质还原反应中,醌类基团涉及的希瓦氏菌的电子传递链是常见的生物化学现象(Newman and Kolter, 2000)。此外,傅里叶变换红外光谱、NMR 光谱和热解-GC-MS 技术的结果也表明醌基是腐殖质中重要的氧化还原活性官能团(Hernández-Montoya et al., 2012; Aeschbacher et al., 2010)。除了醌类官能团之外,非醌芳族结构、含硫官能团和络合金属离子也被认为是腐殖质中潜在的氧化还原活性官能团(Struyk and Sposito, 2001; Einsiedl et al., 2008)。

目前为止,大多数研究一直关注溶解态腐殖质的电子穿梭。溶解态腐殖质的氧化还原活性官能团可以在溶液中较为充分地展开，导致它们易于被微生物接近并因此发挥最大的电子穿梭功能。然而，天然土壤中的腐殖质由于与矿物表面产生多种相互作用而以固态或颗粒态形式存在（Stevenson，1994；Kleber et al.，2007）。这种矿物-有机物的络合结构与微生物难以接近固相腐殖质的过程紧密相关（Torn et al.，1997）。一个新的模型表明微生物接近基质是一个调节天然土壤中有机物功能的过程（Schmidt et al.，2011；Dungait et al.，2012；Lehmann and Kleber，2015）。因此，推测天然土壤中的矿物-有机物络合结构可以限制微生物接近腐殖质的氧化还原活性官能团，从而影响固相腐殖质的电子穿梭。土壤中有机质的保护机制包括了形成具有隔离基质和聚合矿物及黏粒功能的团聚体（Six et al.，2000，2002）。这种保护机制可以使土壤基质内微生物附近的有机物质减少（Davidson and Janssens，2006；Dungait et al.，2011）。为了让土壤粒级达到微生物可利用的程度，颗粒有机物质可以通过基于粒度的物理分馏来分离。不同的粒度代表不同的团聚体水平和不同的对有机物质的物理保护程度，从而代表了天然土壤中不同的微生物可利用度（Six et al.，2002；von Lützow et al.，2007）。因此，不同土壤物理组分中的固相腐殖质也可能表现出不同的电子穿梭活性。

考虑到上述信息，可以认为土壤中固相腐殖质的电子穿梭机制与从土壤中提取的溶解态腐殖质不同。虽然以前的工作提供了在沉积物中的固态腐殖质在微生物还原反应中发生胞外电子穿梭的证据（Roden et al.，2010），但很少研究固相腐殖质电子穿梭及其在天然土壤中微生物可利用性之间的联系。显然，这些联系值得研究，因为可以阐明与土壤腐殖质氧化还原功能有关的环境生物地球化学及污染物转化过程的机制。

在这项工作中，我们旨在研究天然土壤中固相腐殖质电子穿梭的几个基本方面。第一，我们测试了大量土壤基质中固相腐殖质的电子穿梭能力是否取决于它们中的氧化还原活性官能团。第二，我们评估了固相腐殖质的微生物可达性对其电子穿梭能力的影响。第三，我们解释了固相腐殖质在不同土壤团聚体中的电子穿梭功能的异质性。第四，进行了介导的电化学测量，以确定全组分土壤及其物理组分中固相腐殖质的微生物可还原程度（MRE），为固相腐殖质的MRE与土壤中的物理化学保护之间的联系提供新的证据。我们希望本书能够提高人们对腐殖质氧化还原性质的理解，及其在土壤生物地球化学和污染物氧化还原反应中的作用。

2.1 土壤原位固相腐殖质电子转移能力——Fe(Ⅲ)还原的证据

2.1.1 土壤固相腐殖质的微生物还原反应中的胞外电子转移过程

基于 Stewart 等（2008）和 Doetterl 等（2015）提出的土壤有机质概念模型，将土壤分为大团聚体组分（250～2000μm）、微团聚体组分（53～250μm）、非团聚体粉砂组分（2～53μm）和非团聚体黏土组分（<2μm），详见图 2-1。

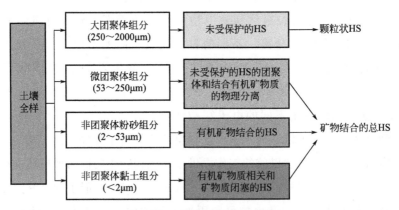

图 2-1 土壤有机质分级方案

土壤预处理程序和电子转移能力的示意如图 2-2 所示。按照图 2-2 所示方法，剥离所有土壤样本中的反应性 Fe，以评估固相腐殖质，将 *Shewanella oneidensis* MR-1 的电子转移到合成的无定形 Fe(Ⅲ) 氧化物的能力。

电子当量并没有在微生物还原和随即过滤后的样品中显著累积，而是在微生物还原但未过滤的样品中按一阶速率定律增加（图 2-3）。这些结果证实了从微生物到土壤固相腐殖质存在直接电子转移，这与之前对沉积物研究的结果一致 (Roden et al., 2010)。在腐殖质提取和未过滤的样品中仅积累了较小的电子当量（图 2-3），表明土壤中不可提取的腐殖质部分和无机基质不具有将微生物中的电子转移到氧化铁(Ⅲ)的能力。在没有经过细胞培养和未过滤的样品中没有显著积累电子当量（图 2-3），意味着固相腐殖质携带的原位电子不负责 Fe(Ⅲ) 的氧化还原。

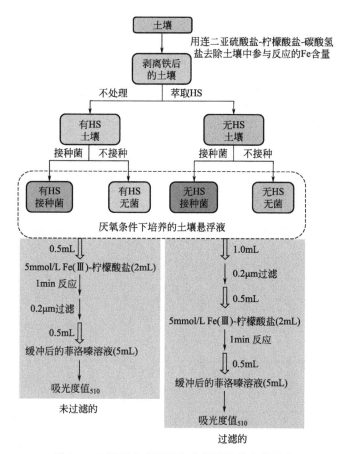

图 2-2 土壤预处理程序和电子转移能力的示意

图 2-3 中,通过电子穿梭能力测定定量转移到固相腐殖质(以电子当量表示)的电子量。虚线表示一阶速率定律的非线性最小二乘回归拟合。

此外,在微生物还原反应实验中,稀盐酸提取的 Fe(Ⅱ) 没有显著累积电子当量(图 2-4),表明了在剥离铁后的土壤和沉积物中缺乏 Fe(Ⅲ) 还原活性(Roden et al. 2010; Lovley and Phillips, 1986)。

图 2-4 中,HCl-Fe(Ⅱ) 是指通过 0.5 mol/L HCl 萃取分析的未过滤样品,然后用邻菲罗啉测定 Fe(Ⅱ)。

2.1.2 土壤腐殖质电子转移能力及其与物理化学性质之间的关系

平行因子分析鉴定了六个独立的荧光组分,包括四种类腐殖质组分(C2、C3、C4 和 C5)、微生物衍生荧光组分(C1)和类蛋白质组分(C6),见图 2-5。

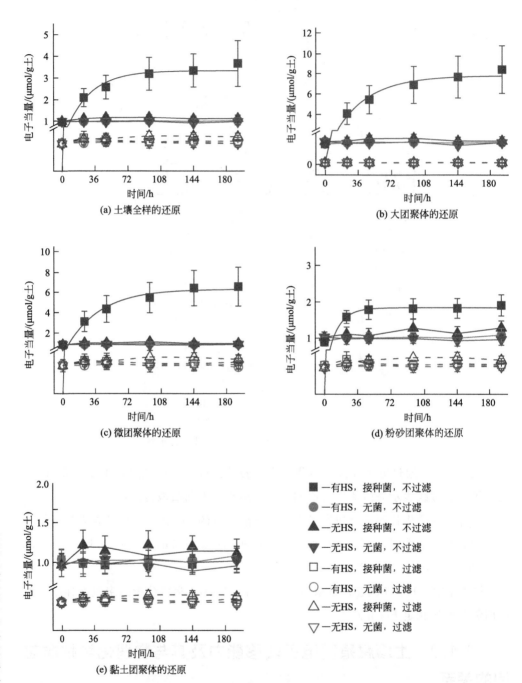

图 2-3 Fe 剥离土壤的微生物还原（一）

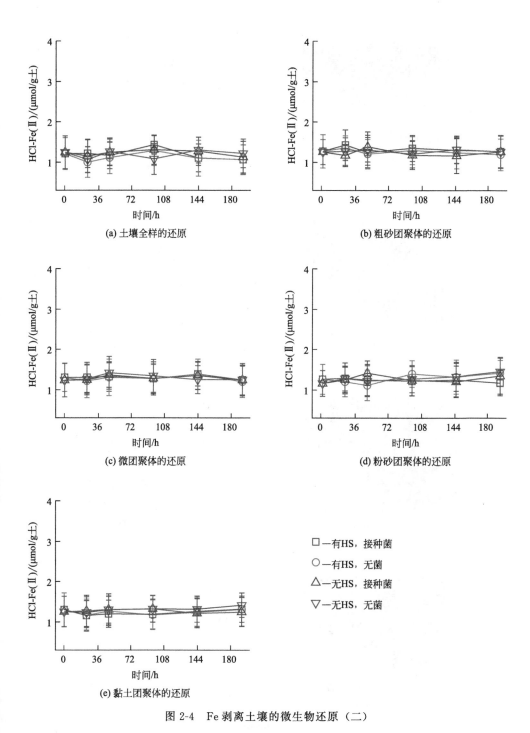

图 2-4　Fe 剥离土壤的微生物还原（二）

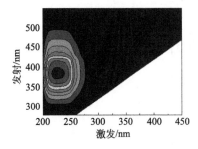
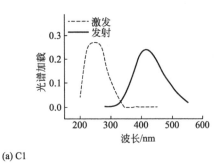

(a) C1

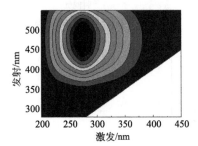
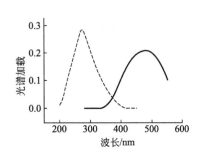

(b) C2

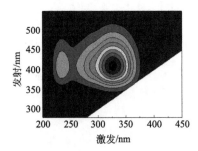
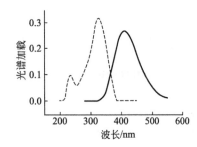

(c) C3

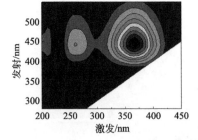
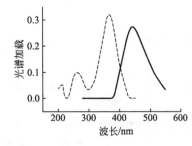

(d) C4

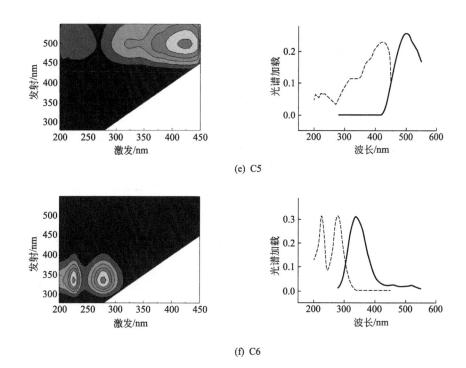

图 2-5　PARAFAC 分析确定的六种组分验证模型的激发和发射负载（见书后彩图 1）

我们采用了许多光学参数（表 2-1）来反映腐殖质的化学结构，以阐明腐殖质的电子穿梭能力与其固有的物理化学性质之间的关系。相关分析表明，在微生物还原反应中，从土壤中转移到溶解态腐殖质的净电子量与 C/H 值呈正相关，而与 E_4/E_6 值呈负相关（图 2-5）。C5 被认定为具有最长发射和激发波长的类腐殖质组分，C6 通常与含 N 化合物有关，也分别对微生物还原反应中转移到溶解态腐殖质的电子净含量产生部分的正负影响（图 2-6）。高 C/H 值通常表示在天然有机物和腐殖质中的芳环高度聚合（Stevenson，1994）。低 E_4/E_6 值主要归因于芳族碳-碳双键官能团的吸收（Kleber and Johnson，2010）。荧光和傅里叶变换离子回旋共振质谱（FT-ICR-MS）分析的直接比较表明，类腐殖质荧光通常与芳族结构共同变化（Herzsprung et al.，2012）。图 2-6 中，转移到溶解性 HS 的电子量以电子当量给出，被用于归一化分析 HS 的质量；HS 的化学结构参数单位如表 2-1 所列，图中数字（省略小数点前的零）显示的颜色表示相关强度；正方形和圆形的背景分别表示正相关和负相关；* 表示在 $p<0.05$ 显著相关。总体而言，我们的研究结果表明，芳香系在溶解态腐殖质中起氧化还原活性基团的作用，与现有观点一致，

醌是溶解态腐殖质中的主要电子接收基团（Lovley et al.，1996；Scott et al.，1998）。

表 2-1　反映腐殖质物理化学性质和土壤有机质变量的指标

变量	说明	单位
元素 C	HS 的元素组成	g/kg HS
元素 H	HS 的元素组成	g/kg HS
元素 O	HS 的元素组成	g/kg HS
元素 N	HS 的元素组成	g/kg HS
元素 S	HS 的元素组成	g/kg HS
元素 C/H	HS 的元素比	—
元素 O/C	HS 的元素比	—
元素 (N+S)/C	HS 的元素比	—
$SUVA_{254}$	254nm 的比紫外线吸收率	L/(m·mg)
E_4/E_6	465nm 和 665nm 的紫外线可见吸收率的比	—
$A_{240\sim400}$	240～400nm 的紫外光谱面积	—
$S_{275\sim295}$	275～295nm 的紫外光谱斜率	—
$S_{350\sim400}$	350～400nm 的紫外光谱斜率	—
HIX	腐殖化指数	—
C1	微生物衍生荧光组分	%
C2	类腐殖质荧光组分	%
C3	类腐殖质荧光组分	%
C4	类腐殖质荧光组分	%
C5	类腐殖质荧光组分	%
C6	类蛋白荧光组分	%
SOM_{tot}	土壤总有机质	g C/kg 土
HS_{tot}	土壤中可提取腐殖质的总含量	g C/kg 土
SPR	特定的潜在呼吸	$\mu g\ C\text{-}CO_2/(g\ 土 \cdot h)$
$SOM_{Microbial}$	微生物可利用土壤有机物	g C/kg 土
$SOM_{Mineral}$	可矿化土壤有机物	g C/kg 土
SOM_{Macro}	土壤大团聚体中的有机质	g C/kg 土

续表

变量	说明	单位
SOM_{Micro}	土壤微团聚体中的有机质	g C/kg 土
$SOM_{Macro+Micro}$	土壤大团聚体和微团聚体中的有机质	g C/kg 土
SOM_S	土壤泥粒中的有机质	g C/kg 土
SOM_C	土壤黏粒中的有机质	g C/kg 土
HS_{Macro}	土壤大团聚体中的可提取腐殖质	g/kg 土
HS_{Micro}	土壤微团聚体中的可提取腐殖质	g/kg 土
$HS_{Macro+Micro}$	土壤大团聚体和微团聚体中的可提取腐殖质	g/kg 土
HS_S	土壤泥粒中的可提取腐殖质	g/kg 土
HS_C	土壤黏粒中的可提取腐殖质	g/kg 土

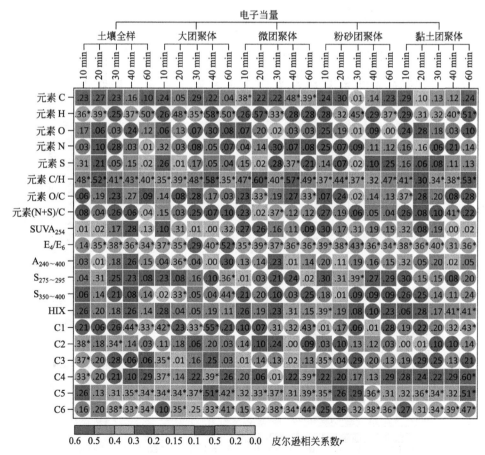

图 2-6 不同微生物转移到溶解性 HS 的电子当量与化学结构参数之间的相关性（见书后彩图 2）

然而，在根际土壤固相腐殖质［图 2-7（a）］中，却没有发现在土壤中提取的溶解态腐殖质中的电子穿梭能力和化学结构之间的这种强相关性。

图 2-7 中，转移到固相 HS 的电子量以电子当量给出，如图 2-7（a）、（b）分别对分析的 HS 和土壤质量进行归一化。HS 和 SOM 变量的化学结构参数单位如表 2-1 所列。所示的颜色和数字（省略小数点前的零）表示相关强度；正方形和圆形的背景分别表示正相关和负相关；* 表示显著相关（$p<0.05$）。

这种现象以及微生物还原相同时间后转移到固相腐殖质的相对于溶解态腐殖质的小电子净量（图 2-8）表明，固相腐殖质中并不是所有的氧化还原活性官能团都能以类似于从土壤中提取的溶解态腐殖质中氧化还原活性官能团的方式将电子从微生物转移到 Fe(Ⅲ) 氧化物。腐殖质中微生物还原反应主要由微生物细胞与腐殖质之间直接接触的电子转移途径所致，尽管微生物可以通过类似于接近 Fe(Ⅲ) 氧化物的几种方式克服接近腐殖质表面的限制（Melton et al.，2014）。因此，我们推测，固相腐殖质中一些氧化还原活性官能团缺乏电子穿梭功能的原因可能是由于这些基团被物理化学保护而难以和微生物直接接触。

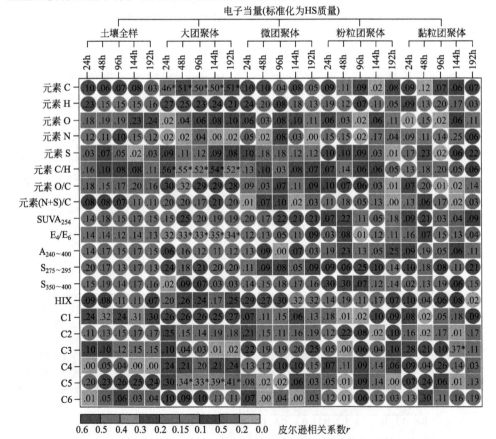

(a) 电子当量与其化学结构参数

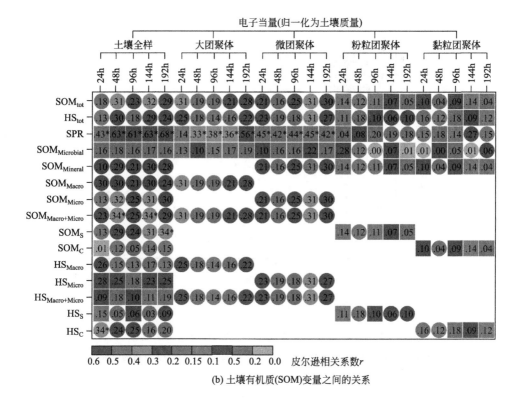

(b) 土壤有机质(SOM)变量之间的关系

图 2-7　不同微生物经一定还原时间后转移到固相 HS 的电子当量与其化学结构参数以及土壤有机质（SOM）变量之间的互相关系（见书后彩图 3）

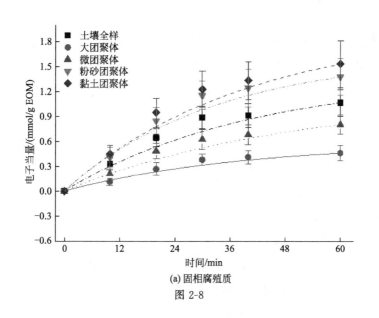

(a) 固相腐殖质

图 2-8

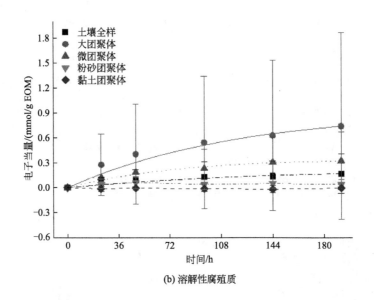

(b) 溶解性腐殖质

图 2-8 微生物还原腐殖质期间所转移的电子量

(转移到固相或溶解性腐殖质的电子量以电子当量给出，其被归一为分析的腐殖质质量)

2.1.3 土壤固相腐殖质的微生物可利用性对其电子穿梭能力的影响

为了验证土壤中固相腐殖质的微生物可及性是否影响电子穿梭能力，我们确定了土壤有机质（SOM）的以下关键变量（Doetterl et al.，2015）：特定潜在呼吸（SPR），微生物可用土壤有机质（SOM$_{微生物}$），矿物相关土壤有机质（SOM$_{矿物}$）和与土壤有机质相关的团聚体组分（表 2-1）。数据显示，分离腐殖质后的土壤中的 SPR 和 SOM$_{微生物}$（图 2-9）显著低于未处理过的土壤（图 2-10）。因此，未处理土壤中的这些变量被认为主要由土壤中可提取碱的腐殖质贡献。

图 2-9 中 SOM 变量单位如表 2-1 所列。随着土壤有机质变化的相同脚注符号在 $p<0.05$ 的土壤分数之间没有显著差异。图 2-10 中 SOM$_{tot}$ 和 HS$_{tot}$ 分别表示土壤有机质和可提取腐殖质的总量；SPR 表示特定潜在呼吸，SOM$_{Microbial}$ 表示微生物可用土壤有机质；SOM$_{Mineral}$ 表示可矿化土壤有机物；SOM$_{Macro}$ 和 HS$_{Macro}$ 分别表示土壤大团聚体中的有机质和可提取腐殖质；SOM$_{Micro}$ 和 HS$_{Micro}$ 分别表示土壤微团聚体中的有机质和可提取腐殖质；SOM$_{Macro+Micro}$ 和 HS$_{Macro+Micro}$ 表示土壤大团聚体

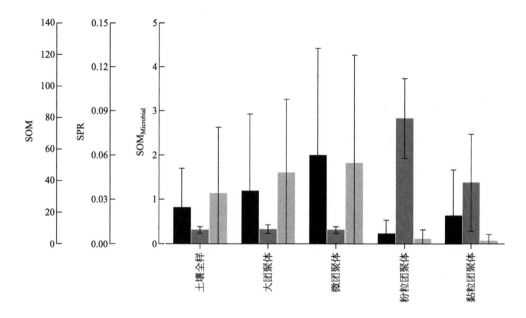

图 2-9 土壤全样和团聚体组分中 SOM、特定潜在呼吸（SPR）和
微生物可用土壤有机质（SOM微生物）

和微团聚体中的有机质和可提取腐殖质；SOM_S 和 HS_S 分别表示土壤泥粒中的有机质和可提取腐殖质；SOM_C 和 HS_C 分别表示土壤黏粒中的有机质和可提取腐殖质。

在土壤微生物还原反应中转移到固相腐殖质的电子净量与 SPR 始终呈正相关 [图 2-7（b）]。高 SPR 值表示高有机物分解速率，这需要微生物与土壤中有机物之间的充分接触。因此，SPR 通常用作微生物对基质可及性的指标（Breulmann et al.，2014；Semenov et al.，2010）。因此，结果表明对固相腐殖质从微生物处接受电子来说与微生物紧密的物理接触是至关重要的。虽然固相腐殖质中的官能团具有氧化还原活性，但由于与微生物之间的空间分离的保护，它们可能难以发挥电子穿梭功能。以前的研究认为，由固相腐殖质或导电细菌纳米导线在沉积物中形成的一个导电网络（Nielsen et al.，2010；Piepenbrock and Kappler，2013），允许电子从一个氧化还原活性基团跳跃至另一个，从而促进更多的氧化还原活性基团实现电子穿梭功能。然而，由于物理化学断裂，这种导电网络可能不会在富含矿物质的土壤中形成，而且即使形成，可能由于腐殖质的氧化还原活性基团之间的距离太长而不允许电子发生跳跃。

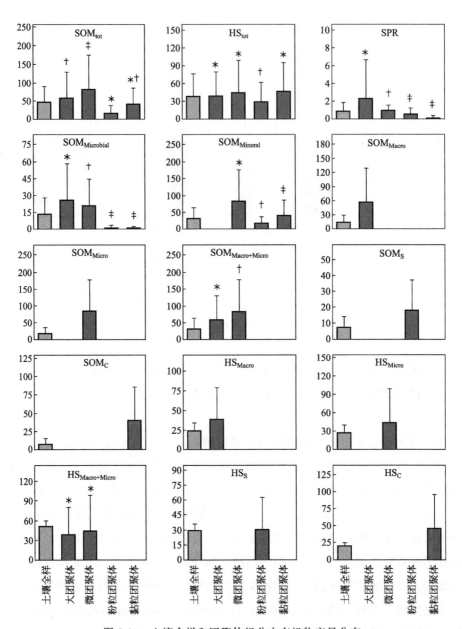

图 2-10 土壤全样和团聚体组分中有机物变量分布

2.1.4 不同土壤团聚体组分中固相腐殖质的电子转移过程的异质性

用土壤物理分离的方法将根际土壤分成几个具有让微生物难以接近有机物质的物理化学屏障的团聚体（von Lützow et al., 2007; Stewart et al., 2008）。我们还对这些组分中固相腐殖质的电子穿梭特性进行了评估，以深入了解土壤中微观的氧化还原环境。我们的研究结果表明，大团聚体组分转移到固相腐殖质的电子净量明显高于小团聚体组分［图2-3（b）～（e）和图2-10］。此外，在微生物还原试验中，电子当量甚至没有显著积累在黏土团聚体组分中［图2-3（e）］。

各种模型已经证明土壤有机质从大团聚体组分转移到小团聚体组分（Six et al., 2000），意味着增强了后者的氧化转化，也因此后者中有机物质的芳香化程度更高。土壤中有机物的这种转移机制可能会导致随着团聚体组分粒径的缩小（图2-11）而引起的C/H值升高和E_4/E_6值降低，从而进一步导致了从大团聚体组分中提取的溶解态腐殖质的电子穿梭能力显著高于提取自小团聚体组分的溶解态腐殖质［图2-8（b）］。然而，与小团聚体组分相比，大团聚体组分中腐殖质的高C/H值和低E_4/E_6值并没有导致其具有更高的电子穿梭能力。因此，固相腐殖质在不同土壤团聚体中的电子穿梭能力的异质性与其氧化还原活性官能团无关。

在小团聚体组分中，具有由包含金属氧化物的有机基团和黏土矿物的相互作用引起的土壤有机质的化学保护，因此其中的腐殖质难以被微生物接近（Six et al., 2002; Puget et al., 2000; Gregorich et al., 2006）。相比之下，无保护机制或轻微的物理隔离是大团聚体组分保护其有机质的方式（Six et al., 2002; Puget et al., 2000），这可能导致其中的腐殖质容易被微生物接近。土壤有机质在团聚体组分之间的独特保护机制支持了我们的结果，即团聚体组分中的SPR和微生物可用性SOM随着团聚体组分尺寸的增加而逐渐降低（图2-11），这可能是导致固相腐殖质在大团聚体组分中拥有较高的电子穿梭能力的原因。

基于SPR指数，认为在大团聚体组分中超过1/2的土壤有机质可被微生物利用，在微生物还原反应中转移到固相腐殖质的净电子量与C/H值和腐殖质中的荧光成分C5呈显著正相关，与腐殖质的E_4/E_6呈负相关［图2-7（a）］；这个发现与从土壤中提取的溶解态腐殖质类似。然而，在其他团聚体组分中却没有发现这种强关联性［图2-7（a）］，其中SOM更多与矿物相关。因此，与溶解性腐殖质不同，固相腐殖质在土壤中的微生物可及性很大程度上是决定其中氧化还原活性官能团数量的关键因素（图2-12），这涉及将电子从微生物转移至Fe(Ⅲ)氧化物，从而将

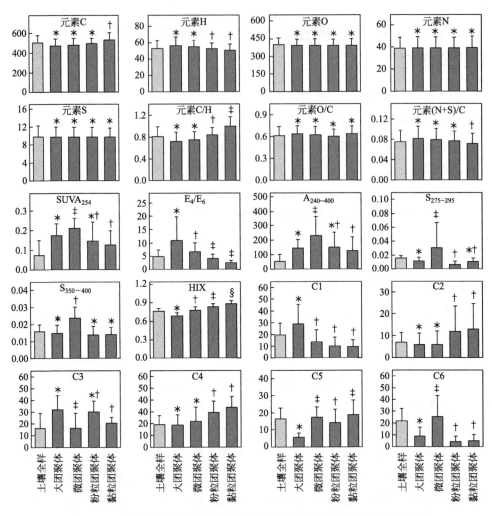

图 2-11 土壤全样和团聚体组分中腐殖质的化学成分结构参数分布情况

注：HS 的化学结构参数单位如表 2-1 所列。随着各个化学指标的相同脚注符号的平均值在 $p<0.05$ 的土壤分数之间没有显著差异

固相腐殖质的电子穿梭与其固有的化学结构联系起来。

在土壤科学中，根据微生物可及性的不同，SOM 可分为活性库、慢性库和惰性库（Parton et al.，1987）。出于有机质封存目的，特别希望增加慢性库和惰性库中的 SOM 总量，以有效减缓二氧化碳从土壤释放到大气中（Amundson，2001）。但为了治理受污染的土壤，在活性库中隔离 SOM，因为 SOM 的高微生物可及性使得固相腐殖质可以在活性库中发挥最大的电子穿梭功能。这将导致固态腐殖质在介导氧化还原有机物和无机污染物参与还原降解以及有力地封锁电子转移到 CO_2，从而降低土壤中的营养性甲烷生成方面发挥重要作用（Blodau and Deppe，2012；

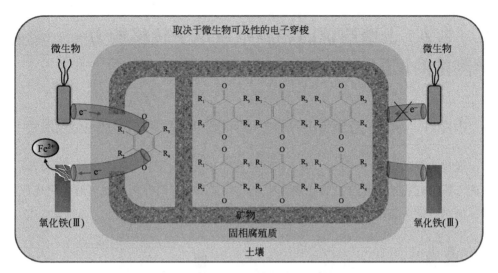

(a) 土壤原位固相腐殖质电子转移机制

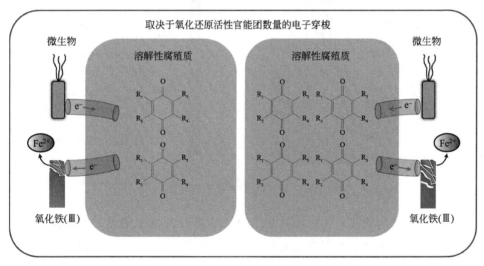

(b) 溶解性腐殖质电子转移机制

图 2-12 腐殖质电子穿梭机制示意

Bridgham et al., 2013；Keller et al., 2009)，从而在保证食品安全的同时减少农业环境危害。鉴于不同库中 SOM 的分布格局在土地利用或管理实践方面是敏感的(Tan et al., 2007)，建议在不同的土地类型中采用不同的土地管理方式，以应对生态环境可持续发展和土壤环境健康的双重挑战。具体来说，在天然的森林和草原土壤中应该保留更多稳定池中的 SOM，因为其土壤污染极少是由人类活动引起，而对于污染比较严重的农业土壤来说，增加高微生物可及性库中的 SOM 似乎更加合理。

2.2 土壤原位固相腐殖质电子转移能力——电化学测量的证据

2.2.1 微生物还原前后土壤中固相腐殖质的氧化还原特性

通过介导的电化学还原（MER）和电化学氧化（MEO）来确定微生物还原前后的土壤（包括土壤全样、各土壤物理组分及其 HS 分离的残留物）的氧化还原状态。测试方法采用改进的电化学方法（Aeschbacher et al.，2010），详见图 2-13。

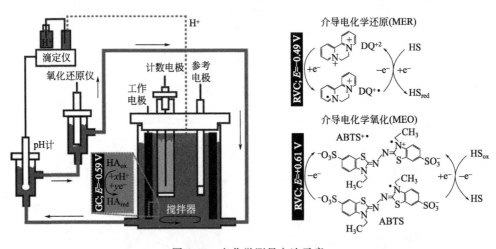

图 2-13　电化学测量方法示意

图 2-13 中玻璃碳瓶不仅充当电化学反应容器，而且还充当工作电极（WE）。根据 Ag/AgCl 参比电极测量所施加的氧化还原电势，并以标准氢电极为参考进行校准。对电极是螺旋铂丝，其通过多孔玻璃粉与工作电极隔开。工作电极钢瓶装有 10mL 缓冲液（0.1mol/L KCl，0.1mol/L 磷酸盐，pH=7），并等于所需电势（即 MER 中 E_h=0.49V，MEO 中 E_h=+0.61V）。将 WE 圆筒填充 10mL 缓冲液（0.1mol/L KCl，0.1mol/L 磷酸盐，pH=7），并平衡至所需电位（E_h=+0.61V）。随后，用 120mL 的电子转移介体 [Diquat dibromide monohydrate (99.5%，Supelco)，DQ]（介导电还原）或 2,2′-联氮-双 (3-乙基苯并噻唑啉-6-磺酸) 的储备溶液（10mmol/L）加入圆筒中，分别产生还原性和氧化性电流峰。最后，在重新获得恒定的背景电流后，将少量土壤悬浮液（100～300mL）添加到圆筒中。通过分别在 MER 和 MEO 中还原和氧化电流响应的积分来量化转移到土壤悬浮液和

从土壤悬浮液转移的电子数量。将积分电流响应标准化为添加土壤中 HS 的质量，以获得 EAC 和 EDC 的值（方程 1 和 2）。

$$\mathrm{EAC} = \frac{\int \frac{I_{\mathrm{red}}}{F} \mathrm{d}t}{m_{\mathrm{HS}}} \tag{1}$$

$$\mathrm{EDC} = \frac{\int \frac{I_{\mathrm{ox}}}{F} \mathrm{d}t}{m_{\mathrm{HS}}} \tag{2}$$

式中，I_{red} 和 I_{ox}（积分区域）分别是经 MER 和 MEO 基准校正的还原电流和氧化电流；m_{HS}（g）是添加土壤中 HS 的质量；F 为法拉第常数，$F=96485\mathrm{C/mol}$。

全土壤中的固相腐殖质含有固有的电子接受能力（EAC）和电子供给能力（EDC），其范围分别为 0.88～1.86mmol（电子）/g（HS）和 0.67～2.32mmol（电子）/g（HS）。缺氧培养 192h 后电子接受能力（EAC）的减少[图 2-14（a）]和电子供给能力（EDC）[图 2-14（b）]的增加可以明显看出，接种 S. oneidensis

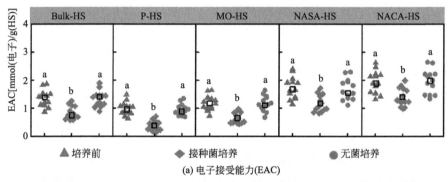

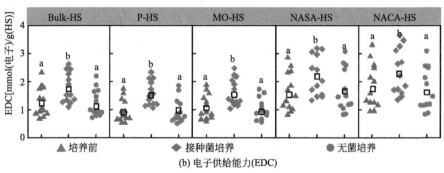

图 2-14　培养前后土壤中接种的和无菌固相腐殖质（HS）的氧化还原特性

注：Bulk-HS 表示土壤全样中的固相腐殖质，P-HS、MO-HS、NASA-HS 和 NACA-HS 分别表示土壤大团聚体组分、微团聚体组分、非团聚体粉砂组分和非团聚体黏土组分中的腐殖质；未填充的正方形表示 EAC 或 EDC 的平均值；相同的小写字母表示无显著差异，不同的小写字母表示显著差异（$p<0.05$）。

MR-1 的土壤全样中固相腐殖质的缺氧培养导致固相腐殖质的微生物显著减少。无细胞接种处理的缺氧培养中土壤全样固相腐殖质的 EAC 和 EDC 没有显著变化 [图 2-14（a）、(b)]。这种情况证实，在缺氧条件下微生物还原导致土壤全样中固相腐殖质的 EAC 和 EDC 发生变化。缺氧微生物培养后，EDC 的增加量约等于 EAC 的减少量（图 2-15），这表明在固相腐殖质的微生物培养过程中，电子接受官能团逐渐转变为电子供给官能团。

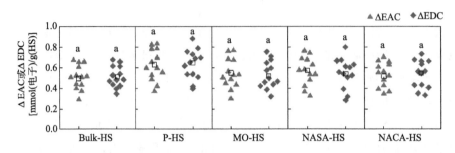

图 2-15　微生物培养后固相腐殖质（HS）的电子接受能力（ΔEAC）的降低和电子供给能力（ΔEDC）的增加

注：Bulk-HS 表示土壤全样中的固相腐殖质，P-HS、MO-HS、NASA-HS 和 NACA-HS 分别表示土壤大团聚体组分、微团聚体组分、非团聚体粉砂组分和非团聚体黏土组分中的腐殖质；未填充的正方形表示 ΔEAC 或 ΔEDC 的平均值；相同的小写字母表示无显著差异，不同的小写字母表示显著差异（$p<0.05$）。

微生物还原程度（MRE）为微生物还原前后固相腐殖质 EAC 的变化百分比，可用来评估腐殖质接受微生物电子的能力。我们的研究结果表明，土壤全样中固相 HS 的 MRE 明显低于 100%（图 2-16），这与以前的观点一致，即由于土壤中有机物的保护机制，并非所有的土壤中固相 HS 氧化还原活性功能部分都可接受微生物电子（Tan et al.，2019）。尽管微生物可以通过各种生存策略最终到达环境固相基质中有机物的官能团（Melton et al.，2014），但在实验阶段土壤中有机物的保护机制可能会严重阻碍微生物与 HS 之间的电子转移。

$$MRE = (EAC_M - EAC_0)/EAC_0$$

式中，EAC_M 为微生物还原后固相腐殖质的 EAC；EAC_0 为微生物还原前固相腐殖质的 EAC。

固相腐殖质的微生物还原程度在不同类型的土壤中表现出显著的波动，范围从 28% 到 47%。在土壤基质中同时采用了各种强度不同的有机质保护机制，以延迟有机质对微生物的可及性（Sollins et al.，1996；Krull et al.，2003；von Lützow et al.，2006）。通过阻塞土壤团聚体、层状硅酸盐或小孔内的有机物对

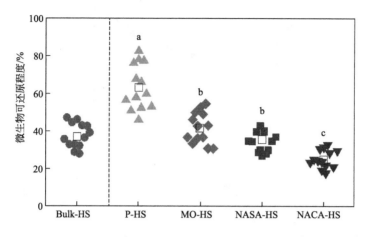

图 2-16 土壤中固相腐殖质的微生物可还原程度

注：Bulk-HS 表示土壤全样中的固相腐殖质，P-HS、MO-HS、NASA-HS 和 NACA-HS 分别表示土壤大团聚体组分、微团聚体组分、非团聚体粉砂组分和非团聚体黏土组分中的腐殖质；未填充的正方形表示微生物可还原程度的平均值；相同的小写字母表示无显著差异，不同的小写字母表示显著差异（$p<0.05$）。

有机物的物理保护以及通过与矿物表面相互作用的化学保护被认为是重要的机制（Sollins et al.，1996；von Lützow et al.，2006；Kelleher and Simpson，2006）。在不同类型的土壤中，不同的有机物保护机制对保护有机物的相对贡献差异很大（Kögel-Knabner et al.，2008；von Lützow et al.，2008）。我们推测这种情况可能使不同土壤中的固相 HS 具有不同的微生物可及性，从而得出不同的微生物还原程度。

如先前的研究（Tan et al.，2019），土壤中固相腐殖质的特定潜在呼吸（SPR）用于进一步验证固相腐殖质的微生物可及性是否会影响土壤中的微生物还原程度。SPR 值高表明有机物分解率高，这要求微生物与土壤中的有机物充分接触。这种情况意味着 SPR 取决于土壤中微生物的生物量、有机物含量和有机物的微生物可及性。因此，通过微生物生物量碳和土壤中腐殖质含量归一化的固相腐殖质的特定潜在呼吸（SPR MBC-HS）可用作微生物对基质可及性的潜在指标（Breulmann et al.，2014；Semenov et al.，2010）。图 2-17（a）描述了土壤全样中固相腐殖质的微生物还原程度与其 SPR MBC-HS 呈显著正相关，支持了以前的观点，即固相腐殖质在土壤中的细胞外电子转移能力取决于微生物的可及性（Tan et al.，2019）。

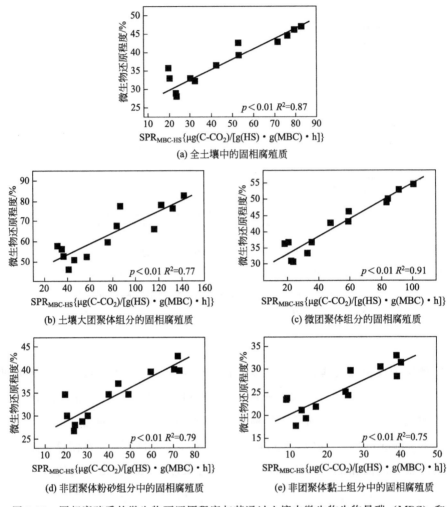

图 2-17 固相腐殖质的微生物可还原程度与其通过土壤中微生物生物量碳（MBC）和 HS 含量（SPR MBC-HS）归一化的潜在呼吸的相关性

2.2.2 不同土壤团聚体中固相腐殖质的氧化还原特性

微生物还原后，P-HS、MO-HS、NASA-HS 和 NACA-HS 的 EAC 和 EDC 分别减少和增加[图 2-14（a）、（b）]，并且 EAC 的减少量几乎等于 EDC 的增加量（图 2-15）。这些结果与土壤全样中固相腐殖质的结果一致。不同土壤团聚体组分的固相腐殖质的微生物还原程度均低于 100%，并且在不同组分之间具有显著差异（图 2-15）。我们的结果证实，P-HS 的 MRE 最高，其次是 MO-HS 和 NASA-HS，而 NACA-HS 的微生物还原程度最低（图 2-16）。在不同的土壤组分中，微生物还原程度的顺序与 SPR MBC-HS 的顺序大致一致（图 2-18）。此外，P-HS、MO-HS、NASA-HS 和 NACA-HS 的微生物还原程度与其相应的 SPR MBC-HS 均呈显著正相关［图 2-17（b）～（e）］。这些结果进一步证实了先前的观点，即尽管不同团聚体组分的固相腐殖质采用了不同的物理化学保护机制，但固相腐殖质的微生物可及性是影响土壤中微生物还原程度的主要因素（Tan et al.，2019）。

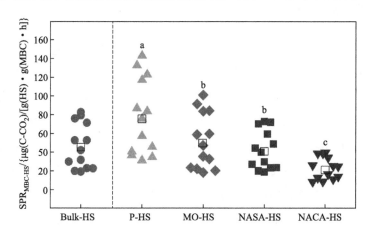

图 2-18 通过微生物生物量碳（MBC）和土壤中 HS 含量（SPR MBC-HS）
归一化的固相腐殖质（HS）的特定潜在呼吸分布

注：Bulk-HS 表示全土壤中的固相腐殖质，P-HS、MO-HS、NASA-HS 和 NACA-HS 分别表示土壤大团聚体组分、微团聚体组分、非团聚体粉砂组分和非团聚体黏土组分中的腐殖质；未填充的正方形表示 SPR MBC-HS 的平均值；相同的小写字母表示无显著差异，不同的小写字母表示显著差异（$p<0.05$）。

P-HS 被认为在土壤中不受保护，因此其微生物可利用性相对较高（Stewart et al.，2008）。这种情况可能是导致 P-HS 的微生物还原程度高于与矿物结合的腐殖质的微生物还原程度的原因（图 2-16）。土壤团聚体形成过程中将有机质包裹到微团聚体中是土壤中最重要的过程之一。将微生物从孔中排除是微团聚体中封闭的土壤有机质的关键保护机制（Sollins et al.，1996）。微团聚体中直径在 0.1～1mm 之间的孔体积急剧增加（McCarthy et al.，2008）。微团聚体中的微环境的尺寸小于微生物的大小，会限制微生物对被封闭的土壤有机质的可及性，从而导致 MO-HS 的 MRE 相对较低（图 2-16）。尽管有机矿物间的相互作用（例如阳离子键、配体交换和范德华力）是非团聚体粉砂中土壤有机质的关键化学保护机制（Kögel-Knabner et al.，2008；Stewart et al.，2008），由于醌类与矿物质表面的相互作用相对较弱，因此相对较容易被微生物攻击，这可能是导致了 NASA-HS 的微生物还原程度未显著低于 MO-HS 的微生物还原程度（图 2-16）。

除了化学保护机制外，层状硅酸盐中的插层也是非聚集黏土组分中土壤有机质的关键物理保护机制（von Lützow et al.，2008）。该特征可能也会降低非聚集黏土组分中固相腐殖质的微生物可及性，从而导致 NACA-HS 的微生物还原程度相对较低（图 2-16）。

2.2.3 环境意义

鉴于固相腐殖质在活性库中发挥了最大的胞外电子转移功能，而对有机物的物理化学保护相对较弱，因此活性库中的腐殖质在介导氧化还原活性的有机和无机污染物方面具有重要作用，通过竞争性抑制电子向 CO_2 的转移来参与还原转化和降低氢营养型甲烷生成（Keller et al.，2009；Blodau and Deppe，2012；Bridgham et al.，2013；Keller and Takagi，2013），从而减少环境风险。我们的结果不仅增进了我们对土壤微环境中腐殖质的原位氧化还原特性的了解，而且为利用土壤有机质用于不同目的的土壤和环境管理提供了理论依据。考虑到土壤有机质的相关部分仍保留在腐殖质分离的土壤残渣中，需要进一步研究土壤中剩余有机质的原位氧化还原特性，以期更好地预测土壤有机质的环境意义。

参考文献

Aeschbacher M, Sander M, Schwarzenbach R P, 2010. Novel electrochemical approach to assess the redox properties of humic substances. Environmental Science & Technology, 44 (1): 87-93.

Amundson R, 2001. The carbon budget in soils. Annual Review of Earth and Planetary Sciences, 29 (1): 535-562.

Benz M, Schink B, Brune A, 1998. Humic acid reduction by Propionibacterium freudenreichii and other fermenting bacteria. Applied and Environmental Microbiology, 64 (11): 4507-4512.

Blodau C, Deppe M, 2012. Humic acid addition lowers methane release in peats of the Mer Bleue bog, Canada. Soil Biology & Biochemistry, 52: 96-98.

Borch T, Kretzschmar R, Kappler A, et al, 2010. Biogeochemical redox processes and their impact on contaminant dynamics. Environmental Science & Technology, 44 (1): 87-93.

Breulmann M, Masyutenko N P, Kogut B M, et al, 2014. Short-term bioavailability of carbon in soil organic matter fractions of different particle sizes and densities in grassland ecosystems. Science of the Total Environment, 497-498: 29-37.

Bridgham S D, Cadillo-Quiroz H, Keller J K, et al, 2013. Methane emissions from wetlands: biogeochemical, microbial, and modeling perspectives from local to global scales. Global Change Biology, 19 (5): 1325-1346.

Cervantes F J, Bok F A M D, Duong-Dac T, et al, 2002. Reduction of humic substances by halorespiring, sulphate-reducing and methanogenic microorganisms. Environmental Microbiology, 4 (1): 51-57.

Davidson E A, Janssens I A, 2006. Temperature sensitivity of soil carbon decomposition and feedbacks to climate change. Nature, 440 (7081): 165-173.

Doetterl S, Stevens A, Six J, et al, 2015. Soil carbon storage controlled by interactions between geochemistry and climate. Nature Geoscience, 8 (10): 780-783.

Dungait J A J, Hopkins D W, Gregory A S, et al, 2012. Soil organic matter turnover is governed by accessibility not recalcitrance. Global Change Biology, 18 (6): 1781-1796.

Dungait J A J, Kemmitt S J, Michallon L, et al, 2011. Variable responses of the soil microbial biomass to trace concentrations of 13C-labelled glucose, using 13C-PLFA analysis. European Journal of Soil Science, 62 (1): 117-126.

Dunnivant F M, Macalady D L, Schwarzenbach R P, 1992. Reduction of substituted nitrobenzenes in aqueous solutions containing natural organic matter. Environmental Science & Technology, 26 (11): 2133-2141.

Einsiedl F, Mayer B, Schäfer T, 2008. Evidence for incorporation of H_2S in groundwater fulvic acids from stable isotope ratios and sulfur K-edge X-ray absorption near edge structure spectroscopy. Environmental Science and Technology, 42 (7): 2439-2444.

Gregorich E G, Beare M H, Mckim U F, et al, 2006. Chemical and biological characteristics of physically uncomplexed organic matter. Soil Science Society of America Journal, 70 (3): 975-985.

Gu B, Chen J, 2003. Enhanced microbial reduction of Cr(Ⅵ) and U(Ⅵ) by different natural organic matter fractions. Geochimica et Cosmochimica Acta, 67 (19): 3575-3582.

Hernández-Montoya V, Alvarez L H, Montes-Morán M A, et al, 2012. Reduction of quinone and non-quinone redox functional groups in different humic acid samples by Geobacter sulfurreducens. Geoderma, 183-184: 25-31.

Herzsprung P, von Tümpling W, Hertkorn N, et al, 2012. Variations of DOM quality in inflows of a drinking water reservoir: Linking of van krevelen diagrams with EEMF spectra by rank correlation. Environmental Science & Technology, 46 (10): 5511-5518.

Kappler A, Haderlein S B, 2003. Natural organic matter as reductant for chlorinated aliphatic pollutants. Environmental Science & Technology, 37 (12): 2714-2719.

Kelleher B P, Simpson A J, 2006. Humic substances in soils: Are they really chemically distinct. Environmental Science & Technology, 40: 4605-4611.

Keller J K, Takagi K K, 2013. Solid-phase organic matter reduction regulates anaerobic decomposition in bog soil. Ecosphere, 4 (5): 1-12.

Keller J K, Weisenhorn P B, Megonigal J P, 2009. Humic acids as electron acceptors in wetland decomposition. Soil Biology & Biochemistry, 41 (7): 1518-1522.

Kleber M, Johnson M G, 2010. Advances in understanding the molecular structure of soil organic matter: Implications for interactions in the environment. Advances in Agronomy, 106: 77-142.

Kleber M, Sollins P, Sutton R, 2007. A conceptual model of organo-mineral interactions in soils: Self-assembly of organic molecular fragments into zonal structures on mineral surfaces. Biogeochemistry, 85 (1): 9-24.

Kögel-Knabner I, Guggenberger G, Kleber M, et al, 2008. Organo-mineral associations in temperate soils: Integrating biology, mineralogy, and organic matter chemistry. Journal of Plant Nutrition and Soil Science, 171 (1): 61-82.

Krull E S, Baldock J A, Skjemstad J O, 2003. Importance of mechanisms and processes of the stabilisation of soil organic matter for modelling carbon turnover. Functional Plant Biology, 30 (2): 207-222.

Lehmann J, Kleber M, 2015. The contentious nature of soil organic matter. Nature, 528 (7580): 60-68.

Lovley D R, Coates J D, BluntHarris E L. et al, 1996. Humic substances as electron acceptors for microbial respiration. Nature, 382 (6590): 445-448.

Lovley D R, Phillips E J P, 1986. Organic matter mineralization with reduction of ferric iron in anaerobic sediments. Applied and Environmental Microbiology, 51 (4): 683-689.

McCarthy J F, Ilavsky J, Jastrow J D, et al, 2008. Protection of organic carbon in soil microaggregates via restructuring of aggregate porosity and filling of pores with accumulating organic matter. Geochimica et Cosmochimica Acta, 72 (19): 4725-4744.

Melton E D, Swanner E D, Behrens S, et al, 2014. The interplay of microbially mediated and abiotic reactions in the biogeochemical Fe cycle. Nature Reviews Microbiology, 12 (12): 797-808.

Nakayasu K, Fukushima M, Sasaki K, 1999. Comparative studies of the reduction behavior of chromium- (Ⅵ) by humic substances and their precursors. Environmental Toxicology and Chemistry, 18 (6): 1085-1090.

Newman D K, Kolter R, 2000. A role of excreted quinones in extracellular electron transfer. Nature, 405 (6782): 94-97.

Nielsen L P, Risgaard-Petersen N, Fossing H, et al, 2010. Electric currents couple spatially separated biogeochemical processes in marine sediment. Nature, 463 (7284): 1071-1074.

Piepenbrock A, Kappler A, 2013. Humic substances and extracellular electron transfer. Microbial Metal Respiration. Microbial Metal Respiration, 107-128.

Parton W J, Schimel D S, Cole C V, et al, 1987. Analysis of factors controlling soil organic matter levels in great plains grasslands. Soil Science Society of America Journal, 51 (5): 1173-1179.

Puget P, Chenu C, Balesdent J, 2000. Dynamics of soil organic matter associated with particle-size fractions of water-stable aggregates. European Journal of Soil Science, 51 (4): 595-605.

Roden E E, Kappler A, Bauer I, et al, 2010. Extracellular electron transfer through microbial reduction of solid-phase humic substances. Nature Geoscience, 3 (6): 417-421.

Schmidt M W I, Torn M S, Abiven S, et al, 2011. Persistence of soil organic matter as an ecosystem property. Nature, 478 (7367): 49-56.

Scott D T, Mcknight D M, Blunt-Harris E L, et al, 1998. Quinone moieties act as electron acceptors in the reduction of humic substances by humics-reducing microorganisms. Environmental Science & Technology, 32 (19): 2984-2989.

Semenov V M, Ivannikova L A, Semenova N A, et al, 2010. Organic matter mineralization in different soil aggregate fractions. Eurasian Soil Science, 43 (2): 141-148.

Six J, Conant R T, Paul E A, et al, 2002. Stabilization mechanisms of soil organic matter: Implications for C-saturation of soils. Plant and Soil, 241 (2): 155-176.

Six J, Elliott E T, Paustian K, 2000. Soil macroaggregate turnover and microaggregate formation: A mechanism for C sequestration under no-tillage agriculture. Soil Biology & Biochemistry, 32 (14): 2099-2103.

Sollins P, Homann P, Caldwell B A, 1996. Stabilization and destabilization of soil organic matter: Mechanisms and controls. Geoderma, 74 (1-2): 65-105.

Stevenson, Frank J, 1994. Humus chemistry: genesis, composition, reactions. John Wiley & Sons.

Stewart C E, Plante A F, Paustian K, et al, 2008. Soil carbon saturation: Linking concept and measurable carbon pools. Soil Science Society of America Journal, 72 (2): 379-392.

Struyk Z, Sposito G, 2001. Redox properties of standard humic acids. Geoderma, 102 (3-4): 329-346.

Tan W, Xi B, Wang G, et al, 2019. Microbial-accessibility-dependent electron shuttling of in situ solid-phase organic matter in soils. Geoderma, 338: 1-4.

Tan Z, Lal R, Owens L, et al, 2007. Distribution of light and heavy fractions of soil organic carbon as related to land use and tillage practice. Soil & Tillage Research, 92 (1-2): 53-59.

Torn M S, Trumbore S E, Chadwick O A, et al, 1997. Mineral control of soil organic carbon storage and turnover. Nature, 389 (6647): 170-173.

Van der Zee F R, Cervantes F J, 2009. Impact and application of electron shuttles on the redox (bio) transformation of contaminants: A review. Biotechnology Advances, 27 (3): 256-277.

von Lützow M, Kögel-Knabner I, Ekschmitt K, et al, 2006. Stabilization of organic matter in temperate soils: Mechanisms and their relevance under different soil conditions-a review. European Journal of Soil Science, 57 (4): 426-445.

von Lützow M, Kögel-Knabner I, Ludwig B, et al, 2008. Stabilization mechanisms of organic matter in four temperate soils: Development and application of a conceptual model. Journal of Plant Nutrition and Soil Science, 171 (1): 111-124.

von Lützow M, Kögel-Knabner I, Ekschmitt K, et al, 2007. SOM fractionation methods: Relevance to functional pools and to stabilization mechanisms. Soil Biology & Biochemistry, 39 (9): 2183-2207.

Wittbrodt P R, Palmer C D, 1997. Reduction of Cr(Ⅵ) by soil humic acids. European Journal of Soil Science, 48 (1): 151-162.

Wolf M, Kappler A, Jiang J, et al, 2009. Effects of humic substances and quinones at low concentrations on ferrihydrite reduction by geobacter metallireducens. Environmental Science & Technology, 43 (15): 5679-5685.

第3章 土壤溶解性腐殖质电子循环能力

由于在腐殖质结构中存在醌-氢醌基团,非醌型芳族结构和络合的金属离子,因此溶解态和固相腐殖质均有氧化还原活性(Scott et al.,1998;Aeschbacher et al.,2011;Struyk and Sposito,2001;Chen et al.,2003;Einsiedl et al.,2008)。在缺氧条件下,腐殖质可以通过不同的微生物还原,如铁还原菌(Lovley et al.,1996)、硫酸盐还原菌(Cervantes et al.,2002)和发酵菌(Benz et al.,1998)。随后,还原的腐殖质可以作为其他价位更高电子受体的电子供体,包括不易接近的 Fe(Ⅲ) 氧化物(Lovley et al.,1996;Bauer and Kappler,2009)、氧(O_2)(Bauer and Kappler,2009) 和各种无机和有机污染物如 U(Ⅵ)(Gu and Chen,2003)、Cr(Ⅵ)(Nakayasu et al.,1999;Wittbrodt and Palmer,1997)、氯化物(Kappler and Haderlein,2003) 和硝基苯(Dunnivant et al.,1992;Van der Zee and Cervantes,2009)。电子从被微生物还原的腐殖质传递到具有氧化还原活性的物质,可以重新氧化被还原的腐殖质,从而恢复腐殖质的原有状态以使其从微生物中持续接受电子(Lovley et al.,1996)。这种所谓的腐殖质的电子穿梭可以极大地影响环境中许多氧化还原活性物质的生物地球化学氧化还原过程。

越来越多的证据表明,腐殖质作为产甲烷条件下的有机末端电子受体可以通过有力地抑制二氧化碳(CO_2)的还原来调节甲烷(CH_4)形成(Miller et al.,2015;Keller et al.,2009;Blodau and Deppe,2012;Bridgham et al.,2013)。在间歇曝气期间,O_2 再氧化经微生物或化学还原的腐殖质时,短时缺氧系统将大大增强这种抑制作用,从而使被氧化的腐殖质恢复还原态,使其在随后的缺氧时段内持续接受电子(Klüpfel et al.,2014)。尽管以前的研究已经为微生物、电化学、化学还原和 O_2 再氧化重复循环中的可逆电子转移提供了证据(Bauer and Kappler,2009;Klüpfel et al.,2014;Ratasuk and Nanny,2007),再氧化后的腐殖质比原始腐殖质还原程度要高，或者需要很长时间来将还原的腐殖质再氧化以将它们的氧化还原状态恢复到与天然腐殖质相同的状态（Bauer and Kappler，2009；Klüpfel et al.，2014）。这意味着并不是腐殖质中所有的氧化还原活性官能团都能通过还原-O_2再氧化的重复循环来穿梭电子。这种现象可能是由于反应溶液中腐殖质的一些被还原的官能团的空间结构经历还原-再氧化诱导的修饰后，避免了被进一步的再氧化（Thieme et al.，2007；Engebretson et al.，1994）。这些修饰程度主要取决于腐殖质的结构柔软性，进一步来说是由腐殖质的固有化学结构所决定的（Von Wandruszka et al.，1999；Duval et al.，2005）。因此，有理由相信在还原-O_2再氧化重复循环中涉及电子穿梭的腐殖质中的氧化还原活性官能团的比例与腐殖质的固有化学结构密切相关。

水稻种植是世界上应用最广泛的粮食作物体系，被认为是重要的人为CH_4来源（Intergovern-mental Panel on Climate Change [IPCC]，2007）。据估计水稻田贡献了全球CH_4排放总量的9%~19%（IPCC，2007；Tokida et al.，2011）。此外，化肥和稻秆在水稻田中的广泛应用也导致了CH_4排放量的增加（Ye et al.，2015；Zhang et al.，2015），加剧了全球变暖。水稻田提供的短时缺氧条件，不仅适于产甲烷，同样也适于腐殖质的氧化还原循环。考虑到CH_4的形成与重复交替缺氧/好氧条件下的腐殖质氧化还原循环密切相关（Miller et al.，2015；Keller et al.，2009；Blodau and Deppe，2012；Bridgham et al.，2013），阐明在微生物还原和O_2再氧化重复循环条件下的电子转移到腐殖质的潜在控制机制，可以更好地了解水稻田土中的厌氧碳循环和CH_4动力学，并且可以发展有广泛前景的农业管理策略，以缓解水稻田中的温室气体排放。

在本书中，从各种水稻田土壤中提取了胡敏酸（HA）和富里酸（FA）两种腐殖质组分。本书旨在量化胡敏酸和富里酸中具有在一个循环（微生物还原和O_2再氧化）中将电子从腐殖质还原性微生物转移到O_2的能力的氧化还原活性官能团的数量，以确定腐殖质的电子穿梭是否可持续地通过连续氧化还原循环，并验证O_2对微生物还原腐殖质的再氧化程度是否与腐殖质的固有化学结构有关。

3.1 厌氧条件下微生物还原后腐殖质的还原能力

根据以前的研究方法，天然腐殖质的还原能力，被微生物还原但未再氧化前的腐殖质，和被微生物还原后立马进行再氧化的腐殖质，都通过介体电化学氧化（MEO）测定（Aeschbacher et al.，2010；Tan et al.，2017）。无菌过滤后，在缺氧条件下（N_2，25℃±1℃），在玻璃碳筒中进行不同氧化还原状态的腐殖质样品的 MEO，作为工作电极（WE）。对 Ag/AgCl 参比电极测量施加的氧化还原电位，参考标准氢电极。对电极是通过多孔玻璃料从 WE 隔室分离的铂丝。将 WE 圆筒填充有 10mL 缓冲液（0.1mol/L KCl，0.1mol/L 磷酸盐，pH=7）并平衡至所需电位（E_h=+0.61V）。随后将 120μL 电子转移介质 2,2'-联氮-双（3-乙基苯并噻唑啉-6-磺酸）的储备溶液（10mmol/L）加入圆筒中，导致氧化电流峰值。在重新获得恒定的背景电流之后，将不同氧化还原状态的少量（即 30~100μL）腐殖质样品掺入圆筒。不同氧化还原状态下 HA（或 FA）亚组分的还原能力测定与 HA（FA）相同。通过以下等式积分氧化电流峰值来量化 RC 值：

$$RC = \frac{\int \frac{I}{F} dt}{m} \tag{1}$$

式中，I（综合区域 A）是 MEO 中基线校正的氧化电流；m 是分析的 HA（或 HA）的质量（g）和分析的 HA（或 FA）亚组分的 C 质量（g C）；F 为法拉第常数，F=96485C/mol。在本书中，微生物还原后经再氧化后的腐殖质（或 HA 和 FA 的亚组分）与天然腐殖质（或 HA 和 FA 的亚组分）的还原能力（RC）之间的差异被称为微生物还原能力（见图 3-1），微生物还原后但尚未再氧化的腐殖质（或 HA 和 FA 的亚组分）的还原能力与经微生物还原和再氧化的腐殖质（或 HA 和 FA 的亚组分）的还原能力的差值的百分比，定义为还原能力的恢复程度（见图 3-1）。

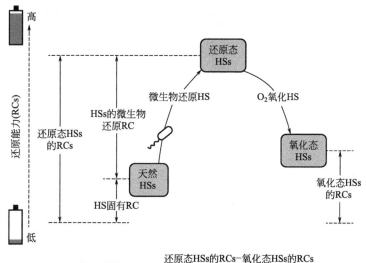

图 3-1 微生物还原腐殖质再氧化过程原理

虽然未还原（天然）腐殖质长时间以干燥形式暴露在 O_2 中，它们含有（187±38）～（217±11）μmol（电子）/g(HS) 的固有还原能力，而胡敏酸和富里酸则分别为（233±12）～（258±30）μmol（电子）/g(HS)。用 *Shewanella putrefaciens* 200 接种的胡敏酸和富里酸的还原能力显著高于其本地的还原能力[见图 3-2（a）、(c)、(e)、(f)]。虽然 *Shewanella oneidensis* MR-1 培育的腐殖质的微生物还原能力与 *Shewanella putrefaciens* 200 培育的腐殖质的微生物还原能力并不完全相等，

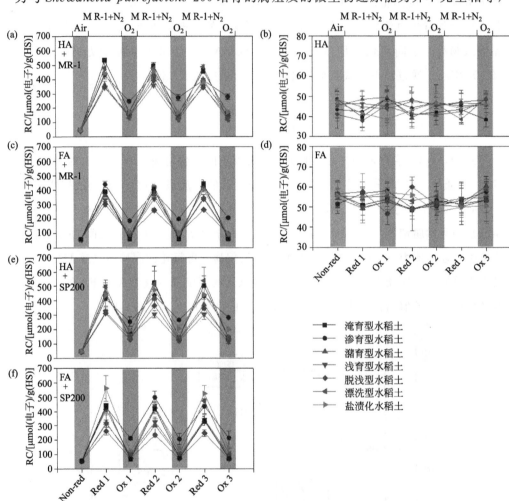

图 3-2　经过间歇好氧厌氧驯化的培育铁还原菌的无菌腐殖质还原能力

注：(a)～(e) 为培育铁还原菌的胡敏酸氧化还原能力 [(a)、(e)]、富里酸 [(c)、(f)]、MR-1 培育 [(a)、(c)]、SP200 培育 [(e)、(f)]、灭菌 HA (b)、灭菌 FA (d) 三次连续的厌氧-好氧循环；Non-red 表示初始状态，Red1、Red2、Red3 分别表示第一次还原、第二次还原、第三次还原，Ox1、Ox2、Ox3 分别表示第一次氧化、第二次氧化、第三次氧化，下同。

但前者与后者具有显著相关性,并且这种良好的相关性对胡敏酸和富里酸两者都适用[见图3-3(b)]。因此,从我们的结果可以推断出,培养系统中腐殖质的相同氧化还原活性基团用于接受来自 MR-1 和 SP200 的电子。

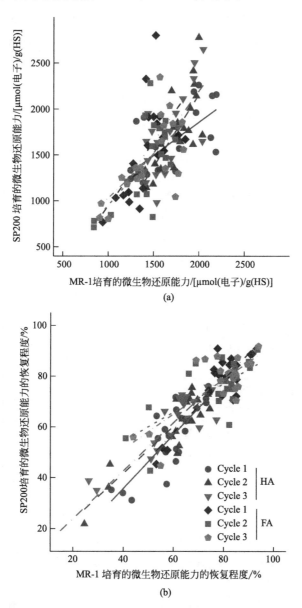

图 3-3　*Shewanella oneidensis* MR-1 培育的微生物还原能力和 *Shewanella putrefaciens* 200 培育的相关性(a);*Shewanella oneidensis* MR-1 培育的微生物还原能力恢复程度和 *Shewanella putrefaciens* 200 培育的相关性(b)

注:HA 和 FA 分别为胡敏酸和富里酸;图中线性显著相关水平($p < 0.05$)。

为了确定腐殖质的微生物还原容易与其固有分子性质之间的关系，我们采用了一组光学参数（见表3-1），以揭示腐殖质的化学结构。包括$S_{250\sim600}$、$A_{240\sim400}$、$SUVA_{254}$、E_4/E_6、HIX、AMW、荧光组分和通过FTIR光谱分析的官能团的相对含量。并行因子分析鉴定了六个独立的荧光组分，包括：四种类腐殖质组分（C1、C2、C3和C4）；与原位产生相关的组分（C5）；类蛋白质组分（C6）（图3-4）。

表3-1 腐殖质化学结构参数总结

变量	说明	单位
元素 C/H	腐殖质的元素比	—
元素 O/C	腐殖质的元素比	—
$SUVA_{254}$	254nm的比紫外线吸收率	L/(m·mg)
E_4/E_6	465nm和665nm的紫外线可见吸收率的比	—
$S_{250\sim600}$	250~600nm的紫外光谱斜率	—
$A_{240\sim400}$	240~400nm的紫外光谱面积	—
HIX	腐殖化指数	—
C1	类腐殖质荧光组分	%
C2	类腐殖质荧光组分	%
C3	类腐殖质荧光组分	%
C4	类腐殖质荧光组分	%
C5	微生物衍生荧光组分	%
C6	类蛋白质荧光组分	%
AMW_n	腐殖质的数均分子量	Dalton
AMW_w	腐殖质的重均分子量	Dalton
FTIR Group 1	腐殖质FTIR光谱中的波数3400~3300cm^{-1}；O—H和N—H伸展	%
FTIR Group 2	腐殖质FTIR光谱中的波数3380cm^{-1}；氢键-羟基	%
FTIR Group 3	腐殖质FTIR光谱中的波数2985cm^{-1}；CH_3和CH_2伸展	%
FTIR Group 4	腐殖质FTIR光谱中的波数2940~2900cm^{-1}；脂肪族C—H拉伸	%
FTIR Group 5	腐殖质FTIR光谱中的波数1725~1720cm^{-1}；C=O延伸的COOH官能团	%
FTIR Group 6	腐殖质FTIR光谱中的波数1650~1630cm^{-1}；C=O延伸，芳族C=C，氢键C=O	%
FTIR Group 7	腐殖质FTIR光谱中的波数1460cm^{-1}；脂肪族C—H，C—CH_3	%

续表

变量	说明	单位
FTIR Group 8	腐殖质 FTIR 光谱中的波数 1440cm^{-1};甲基的 C—H 链	%
FTIR Group 9	腐殖质 FTIR 光谱中的波数 1435cm^{-1};C—H 弯曲	%
FTIR Group 10	腐殖质 FTIR 光谱中的波数 1400cm^{-1};COO$^-$ 反对称延伸	%
FTIR Group 11	腐殖质 FTIR 光谱中的波数 1390cm^{-1};COOH 盐	%
FTIR Group 12	腐殖质 FTIR 光谱中的波数 1280～1230cm^{-1};C—O 延伸,芳族 C—O、C—O 酯键	%
FTIR Group 13	腐殖质 FTIR 光谱中的波数 1170～950cm^{-1};糖苷键典型的 C—C、C—OH、C—O—C,多糖 C—O 延伸	%
FTIR Group 14	腐殖质 FTIR 光谱中的波数 1035cm^{-1};O—CH$_3$ 振动	%
FTIR Group 15	腐殖质 FTIR 光谱中的波数 840cm^{-1};芳香族 C—H 振动	%

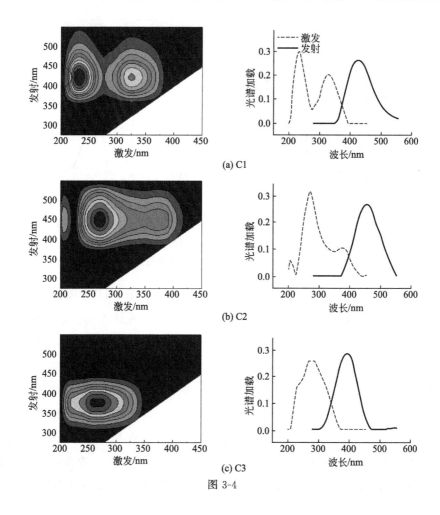

图 3-4

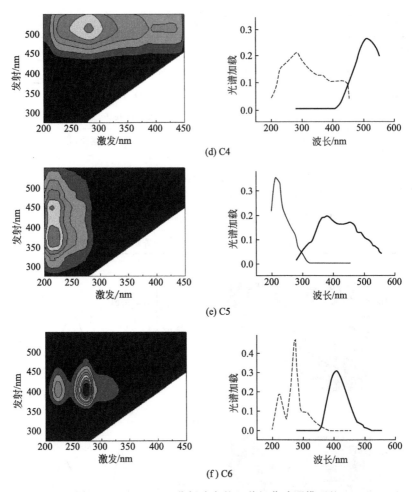

图 3-4 PARAFAC 分析确定的六种组分验证模型的
激发和发射负载（见书后彩图 4）

相关分析表明，腐殖质的微生物还原能力与 C/H 值、HIX 和类腐殖质组分（C4 或 C6）呈正相关，与 E_4/E_6 和组分 C5 呈负相关（见图 3-5）。值得注意的是，无论胡敏酸和富里酸的独立数据集还是胡敏酸和富里酸的集成数据集（见图 3-5），腐殖质的微生物还原能力与其相关化学结构之间的关系是一致的，表明与腐殖质的微生物还原能力相关的化学结构独立于腐殖质组分。腐殖质的高 C/H 值和 HIX 值通常表示高度缩合的芳环（Stevenson，1994；Ohno，2002）。傅里叶变换离子回旋共振质谱（FT-ICR-MS）和荧光分析的直接比较表明，类腐殖质荧光通常与芳族结构共同变化（Herzsprung et al.，2012）。低 E_4/E_6 值主要是由于芳香族碳-碳双键官能团的吸收（Kleber and Johnson，2010）。上述信息表明，腐殖质的 C/H 值、HIX、

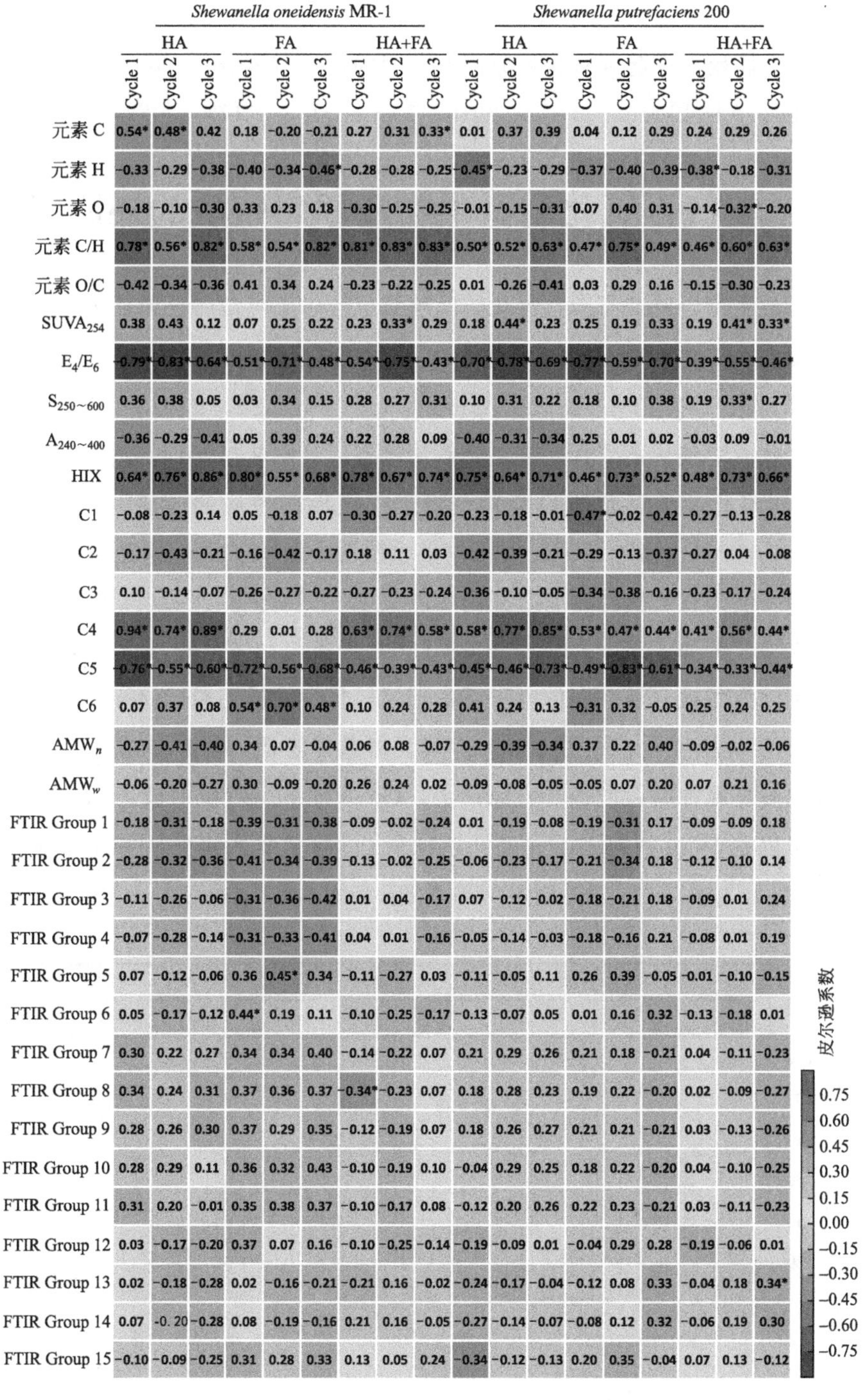

图 3-5　微生物还原能力和表征腐殖质化学结构的参数相关性（见书后彩图 5）

注：HA+FA 表示统计分析中胡敏酸和富里酸的集合数据集；红外官能团 1~15 表示通过红外光谱分析出的功能性基团；颜色和数字表示相关性的正负性和强度；（*）表示显著相关（$p<0.05$）；Cycle1、Cycle2、Cycle3 分别表示第一次循环、第二次循环、第三次循环；下同。

类腐殖质组分和 E_4/E_6 可用于表示醌类结构。因此，我们的研究结果表明，芳烃体系如醌类在腐殖质中起到氧化还原活性基团的作用（Scott et al.，1998；Lovley et al.，1996；Aeschbacher et al.，2010；Tan et al.，2017）。

3.2 氧气再氧化后微生物还原腐殖质的还原能力

虽然经微生物还原但尚未再氧化的腐殖质也产生了可观的还原能力，但还原后的腐殖质暴露在 O_2 中 1h 后，胡敏酸和富里酸的还原能力分别下降到（759±185）μmol（电子）/g(HS) 和（464±163）μmol（电子）/g(HS)。该结果表明，O_2 能够潜在地将腐殖质溶液从还原状态恢复到氧化状态 [见图 3-3（a）]。然而，在微生物还原和 O_2 再氧化后的腐殖质仍然显著高于未还原腐殖质的还原能力 [见图 3-3（a）、（c）、（e）、（f）和图 3-6]，表明由于腐殖质中某些被微生物还原后的官能团具有持续性，使其在实验的时间范围内不能被氧气再氧化。这一观察结果也得到先前研究的支持。例如，Bauer 和 Kappler（2009）指出，还原和氧气再氧化后的腐殖质转移到柠檬酸铁（Ⅲ）中的电子，是不经过化学物质还原但储存在空气中几个月的腐殖质的 4 倍。Klüpfel 等（2014）观察到微生物还原的腐殖质在再氧化后比在实验开始时还原性略微增强。已经证明，O_2 条件下腐殖质中被还原官能团的再氧化过程可以在 1min 内完成，并且经还原和 O_2 再氧化的腐殖质的还原能力不会随着暴露在 O_2 中时间的延长而进一步降低（Bauer and Kappler，2009）。因此，

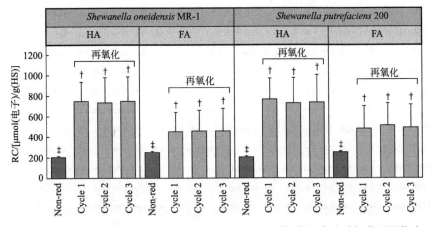

图 3-6　比较未还原的腐殖质和在微生物还原后 O_2 再氧化的腐殖质氧化还原能力

注：相同脚注的胡敏酸或富里酸的平均值（$n=21$）显著相关水平 $p<0.05$。

通过 O_2 再氧化的微生物氧化还原活性官能团的过程，在 1h 内就可能完成。

虽然腐殖质是高分子有机化合物，但是当它们溶解在溶液中时，由于它们分子形状的特征为柔性线性、折叠链或无规则卷曲结构（Tan，2014），它们在空间结构方面表现出潜在的可变性（Von Wandruszka et al.，1999；Maurice and Namjesnik-Dejanovic，1999；Myneni et al.，1999）。因此，溶液中腐殖质的还原和再氧化可能导致其空间结构的重组。这一假设也得到其他研究的支持。例如，Engebretson 和 Von Wandruszka（1994）指出，腐殖质中笼状和假胶束结构的存在可保护荧光团免受溴化物淬灭。Thieme 等（2007）通过使用透射 X 射线显微镜演示了这些物质在经历氧化还原反应后腐殖质的形态学变化。考虑到腐殖质空间结构的变化可能通过改变腐殖质的暴露面而潜在地限制腐殖质中 O_2 扩散微孔的可及性，从而抑制被还原的腐殖质官能团的进一步再氧化的反应活性（Schwarzenbach et al.，2005）。我们推断，我们的假设可以解释为什么当腐殖质溶解在溶液中时，腐殖质结构中所有被微生物还原的氧化还原活性官能团都不能被 O_2 迅速地再氧化。

我们计算了还原能力的恢复程度，以评估被微生物还原的腐殖质在 O_2 刺激下的再氧化。方差分析显示，腐殖质样本之间胡敏酸与富里酸的还原能力恢复程度有显著性差异（$p<0.05$）。相比之下，MR-1 和 SP200 之间的还原能力恢复程度并没有显著性差异（$p>0.05$）（见表 3-2）。此外，MR-1 的培育系统中还原能力的恢复程度与 SP200 培育系统中的恢复程度显著相关［见图 3-3（b）］。因此，我们的研究结果表明，还原能力的恢复程度与铁还原菌无关，而与腐殖质的组分和来源有关。尽管类似的假定结构，特征分子量、元素组成和官能团含量可能在不同来源和组分的腐殖质样品之间有所差异。因此，我们推测还原能力的恢复程度可能与腐殖质的固有化学结构有关。如图 3-7 所示，无论是胡敏酸和富里酸的单个数据集还是胡敏酸和富里酸的集成数据集，还原能力的恢复程度都与腐殖质的数均分子量（AMW_n）和重均分子量（AMW_w）呈显著负相关。此外，当将胡敏酸和富里酸样品进一步分离成两个具有差异分子量的亚组分时，结果表明，具有低分子量的胡敏酸（或富里酸）亚组分的还原能力恢复程度明显高于高分子量的（见图 3-8）。该观察结果适用于所有胡敏酸和富里酸样品（见图 3-9 和图 3-10）。总之，这些研究结果表明，具有较大的平均分子量比具有较小平均分子量的微生物还原腐殖质更难通过氧气从还原状态恢复到氧化状态。

表 3-2 微生物、氧化还原循环、腐殖质组分（胡敏酸和富里酸）、腐殖质样品对微生物氧化还原能力、还原能力恢复程度和氧化还原反应电子转移量的影响

变量			微生物 (A)	氧化还原循环(B)	HS组分 (C)	HS样品 (D)	A×B	A×C	A×D	B×C	B×D	C×D	A×B×C	A×B×D	A×C×D	B×C×D	A×B×C×D
微生物 RC	微生物	MR-1		NS	*	*				NS	NS	NS				NS	
		SP200	NS	NS	*	*				NS	NS	NS				NS	
	氧化还原循环	1	NS		*	*		NS	NS			NS		NS	NS		
		2	NS		*	*		NS	NS			NS		NS	NS		
		3	NS		*	*		NS	NS					NS	NS		
	HS组分	HA	NS	NS		*	NS			NS	NS		NS			NS	
		FA	NS	NS		*	NS			NS	NS		NS			NS	
	HS样品	1	NS	NS	*		NS	NS		NS	NS		NS				
		2	NS	NS	*		NS	NS		NS	NS		NS				
		3	NS	NS	*		NS	NS		NS	NS		NS				
		4	NS	NS	*		NS	NS		NS	NS		NS				
		5	NS	NS	*		NS	NS		NS	NS		NS				
		6	NS	NS	*		NS	NS		NS	NS		NS				
		7	NS	NS	*		NS	NS		NS	NS		NS				
RC的恢复程度	微生物	MR-1		NS	*	*				NS	NS	NS				NS	
		SP200	NS	NS	*	*				NS	NS	NS				NS	
	氧化还原循环	1	NS		*	*		NS	NS			NS		NS	NS		
		2	NS		*	*		NS	NS			NS		NS	NS		
		3	NS		*	*		NS	NS			NS		NS	NS		
	HS组分	HA	NS	NS		*	NS			NS	NS		NS			NS	
		FA	NS	NS		*	NS			NS	NS		NS			NS	

续表

变量		因子														
		微生物(A)	氧化还原循环(B)	HS组分(C)	HS样品(D)	A×B	A×C	A×D	B×C	B×D	C×D	A×B×C	A×B×D	A×C×D	B×C×D	A×B×C×D
RC的恢复程度	HS样品 1	NS	NS	*		NS	NS	NS	NS			NS				
	2	NS	NS	*		NS	NS	NS	NS			NS				
	3	NS		*	*	NS	NS	NS	NS			NS				
	4	NS	NS	*	*	NS	NS	NS	NS			NS				
	5	NS	NS	*		NS	NS	NS	NS			NS				
	6	NS	NS	*	*	NS	NS	NS	NS			NS				
	7	NS	NS	NS		NS	NS	NS	NS			NS				
	微生物 MR-1									NS	NS	NS			NS	
	SP200		NS	NS					NS	NS	NS			NS		
	氧化还原循环 1	NS		NS	*	NS	NS	NS		NS						
	2	NS		NS	*	NS	NS	NS		NS			NS			
	3	NS		NS	*	NS	NS	NS		NS			NS			
氧化还原循环中涉及电子数	HS组分 HA	NS	NS		*	NS	NS	NS	NS	NS			NS			
	FA	NS	NS		*	NS	NS	NS	NS	NS			NS			
	HS样品 1	NS	NS	NS		NS	NS	NS	NS	NS	NS	NS				
	2	NS	NS	NS		NS	NS	NS	NS	NS	NS	NS				
	3	NS	NS	NS		NS	NS	NS	NS	NS	NS	NS				
	4	NS	NS	NS		NS	NS	NS	NS	NS	NS	NS				
	5	NS	NS	NS		NS	NS	NS	NS	NS	NS	NS				
	6	NS	NS	NS		NS	NS	NS	NS	NS	NS	NS				
	7	NS	NS	NS		NS	NS	NS	NS	NS	NS	NS				

注：（*）表示统计学上显著相关（$p<0.05$）；NS 表示统计学上没有显著相关（$p>0.05$）。

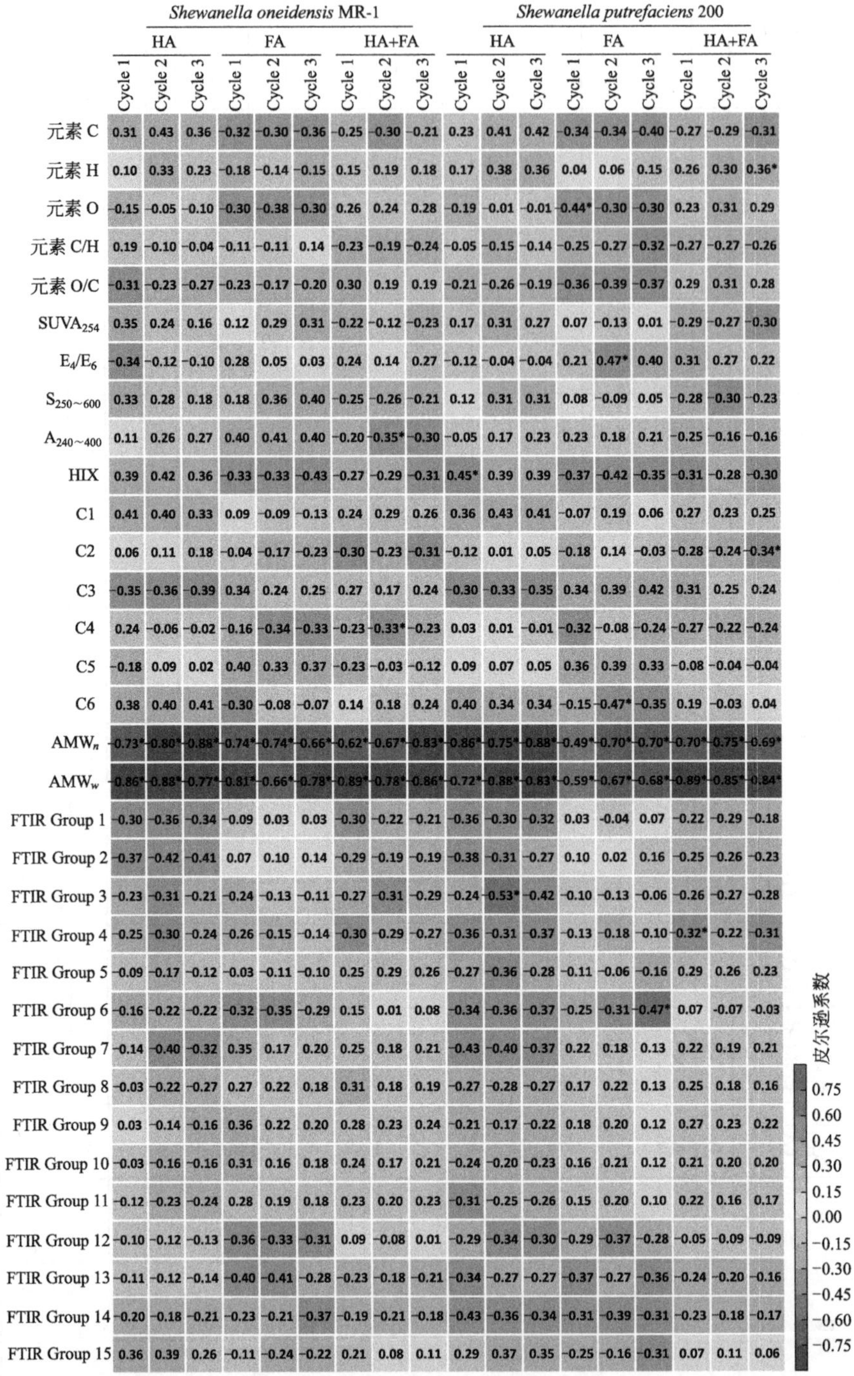

| | Shewanella oneidensis MR-1 ||||||||| Shewanella putrefaciens 200 |||||||||
| | HA ||| FA ||| HA+FA ||| HA ||| FA ||| HA+FA |||
	Cycle 1	Cycle 2	Cycle 3	Cycle 1	Cycle 2	Cycle 3	Cycle 1	Cycle 2	Cycle 3	Cycle 1	Cycle 2	Cycle 3	Cycle 1	Cycle 2	Cycle 3	Cycle 1	Cycle 2	Cycle 3
元素 C	0.31	0.43	0.36	-0.32	-0.30	-0.36	-0.25	-0.30	-0.21	0.23	0.41	0.42	-0.34	-0.34	-0.40	-0.27	-0.29	-0.31
元素 H	0.10	0.33	0.23	-0.18	-0.14	-0.15	0.15	0.19	0.18	0.17	0.38	0.36	0.04	0.06	0.15	0.26	0.30	0.36*
元素 O	-0.15	-0.05	-0.10	-0.30	-0.38	-0.30	0.26	0.24	0.28	-0.19	-0.01	-0.01	-0.44*	-0.30	-0.30	0.23	0.31	0.29
元素 C/H	0.19	-0.10	-0.04	-0.11	-0.11	0.14	-0.23	-0.24	-0.24	-0.05	-0.15	-0.14	-0.25	-0.27	-0.32	-0.27	-0.27	-0.26
元素 O/C	-0.31	-0.23	-0.27	-0.23	-0.17	-0.20	0.30	0.18	0.19	-0.21	-0.26	-0.19	-0.36	-0.39	-0.37	0.29	0.31	0.28
$SUVA_{254}$	0.35	0.24	0.16	0.12	0.29	0.31	-0.22	-0.12	-0.23	0.17	0.31	0.27	0.07	-0.13	0.01	-0.29	-0.27	-0.30
E_4/E_6	-0.34	-0.12	-0.10	0.28	0.05	0.03	0.24	0.14	0.27	-0.12	-0.04	-0.04	0.21	0.47*	0.40	0.31	0.27	0.22
$S_{250\sim600}$	0.33	0.28	0.18	0.18	0.36	0.40	-0.25	-0.26	-0.21	0.12	0.31	0.31	0.08	-0.09	0.05	-0.28	-0.30	-0.23
$A_{240\sim400}$	0.11	0.26	0.27	0.40	0.41	0.40	-0.20	-0.35*	-0.30	-0.05	0.17	0.23	0.23	0.18	0.21	-0.25	-0.16	-0.16
HIX	0.39	0.42	0.36	-0.33	-0.33	-0.43	-0.27	-0.29	-0.31	0.45*	0.39	0.39	-0.37	-0.42	-0.35	-0.31	-0.28	-0.30
C1	0.41	0.40	0.33	0.09	-0.09	-0.13	0.24	0.29	0.26	0.36	0.43	0.41	-0.07	0.19	0.06	0.27	0.23	0.25
C2	0.06	0.11	0.18	-0.04	-0.17	-0.23	-0.30	-0.23	-0.31	-0.12	0.01	0.05	-0.18	0.14	-0.03	-0.28	-0.24	-0.34*
C3	-0.35	-0.36	-0.39	0.34	0.24	0.23	0.27	0.17	0.24	-0.30	-0.33	-0.35	0.34	0.39	0.42	0.31	0.25	0.24
C4	0.24	-0.06	-0.02	-0.16	-0.34	-0.33	-0.33*	-0.23	0.03	0.01	-0.01	-0.32	-0.06	-0.24	-0.27	-0.22	-0.24	
C5	-0.18	0.09	0.02	0.40	0.33	0.37	-0.23	-0.03	-0.12	0.09	0.07	0.05	0.36	0.39	0.33	-0.08	-0.04	-0.04
C6	0.38	0.40	0.41	-0.30	-0.08	-0.07	0.14	0.38	0.40	0.34	0.34	-0.15	-0.47*	-0.35	0.19	-0.03	0.04	
AMW_n	-0.73*	-0.80*	-0.88*	-0.74*	-0.74*	-0.66*	-0.62*	-0.67*	-0.83*	-0.86*	-0.75*	-0.88*	-0.49*	-0.70*	-0.70*	-0.70*	-0.75*	-0.69*
AMW_w	-0.86*	-0.88*	-0.77*	-0.81*	-0.66*	-0.78*	-0.89*	-0.78*	-0.86*	-0.72*	-0.88*	-0.83*	-0.59*	-0.67*	-0.68*	-0.89*	-0.85*	-0.84*
FTIR Group 1	-0.30	-0.36	-0.34	-0.09	0.03	0.03	-0.30	-0.22	-0.21	-0.30	-0.30	-0.32	-0.04	0.07	-0.36	-0.22	-0.29	-0.18
FTIR Group 2	-0.37	-0.42	-0.41	0.07	0.10	0.14	-0.29	-0.19	-0.19	-0.38	-0.31	-0.27	0.10	0.02	0.16	-0.25	-0.26	-0.23
FTIR Group 3	-0.23	-0.31	-0.21	-0.24	-0.13	-0.11	-0.27	-0.31	-0.29	-0.24	-0.53*	-0.42	-0.10	-0.13	-0.06	-0.26	-0.27	-0.28
FTIR Group 4	-0.25	-0.30	-0.24	-0.26	-0.15	-0.14	-0.30	-0.29	-0.27	-0.31	-0.37	-0.35	-0.13	-0.18	-0.10	-0.32*	-0.22	-0.31
FTIR Group 5	-0.09	-0.17	-0.12	-0.03	-0.11	-0.10	0.25	0.29	0.26	-0.27	-0.36	-0.28	-0.11	-0.06	-0.16	0.29	0.26	0.23
FTIR Group 6	-0.16	-0.22	-0.22	-0.32	-0.35	-0.29	0.15	0.01	0.08	-0.34	-0.36	-0.37	-0.25	-0.31	-0.47*	0.07	-0.07	-0.03
FTIR Group 7	-0.14	-0.40	-0.32	0.35	0.17	0.20	0.25	0.18	0.21	-0.43	-0.40	-0.37	0.18	0.13	0.22	0.17		
FTIR Group 8	-0.03	-0.22	-0.27	0.27	0.22	0.18	0.31	0.18	0.19	-0.27	-0.28	-0.27	0.17	0.22	0.13	0.25	0.18	0.16
FTIR Group 9	0.03	-0.14	-0.16	0.36	0.22	0.20	0.28	0.23	0.24	-0.21	-0.22	-0.22	0.18	0.20	0.12	0.27	0.23	0.22
FTIR Group 10	-0.03	-0.16	-0.16	0.31	0.18	0.24	0.24	0.17	0.21	-0.24	-0.28	-0.26	0.16	0.20	0.12	0.21	0.25	0.22
FTIR Group 11	-0.12	-0.23	-0.24	0.28	0.19	0.18	0.23	0.20	0.23	-0.31	-0.25	-0.26	0.15	0.20	0.10	0.22	0.16	0.17
FTIR Group 12	-0.10	-0.12	-0.13	-0.36	-0.33	-0.31	0.09	-0.08	0.01	-0.26	-0.30	-0.30	-0.37	-0.28	-0.05	-0.09	-0.09	
FTIR Group 13	-0.11	-0.12	-0.14	-0.40	-0.41	-0.34	-0.23	-0.18	-0.21	-0.34	-0.27	-0.37	-0.27	-0.36	-0.24	-0.20	-0.16	
FTIR Group 14	-0.20	-0.18	-0.21	-0.23	-0.21	-0.37	-0.19	-0.21	-0.18	-0.43	-0.36	-0.34	-0.31	-0.39	-0.31	-0.23	-0.18	-0.17
FTIR Group 15	0.36	0.39	0.26	-0.11	-0.24	-0.22	0.21	0.08	0.11	0.29	0.37	0.35	-0.25	-0.16	-0.31	0.07	0.11	0.06

图 3-7 还原能力恢复程度和表征腐殖质化学结构的参数相关性（见书后彩图 6）

注：HA+FA 表示统计分析中胡敏酸和富里酸的集合数据；红外官能团 1~15 表示通过红外光谱分析出的功能性基团；颜色和数字表示相关性的正负性和强度；（*）表示在 $p<0.05$ 显著相关。

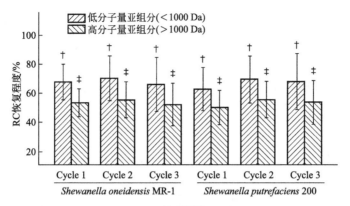

(a) 胡敏酸

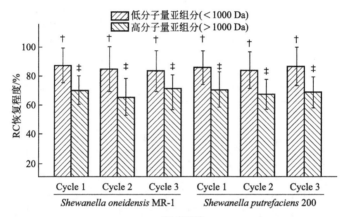

(b) 富里酸

图 3-8　比较氧化还原循环中低分子量（<1000 Da）亚组分和高分子量（>1000Da）亚组分氧化还原能力恢复程度

注：相同脚注的胡敏酸或富里酸的平均值（$n=21$）显著相关（$p<0.05$）；1Da＝1g/mol，下同。

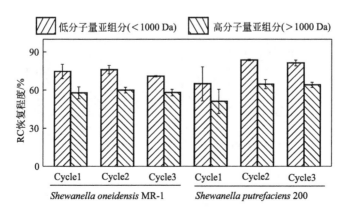

(a) 淹育型水稻土

图 3-9

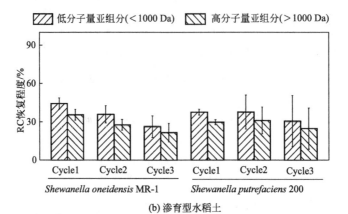

(b) 渗育型水稻土

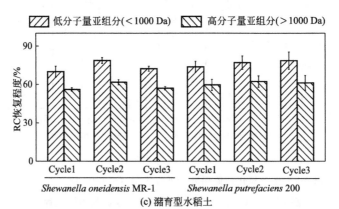

(c) 潴育型水稻土

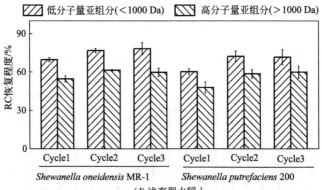

(d) 浅育型水稻土

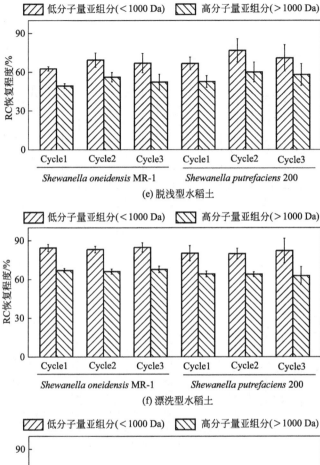

图 3-9 比较胡敏酸低分子量（<1000Da）亚组分和
高分子量（>1000Da）亚组分氧化还原能力恢复程度

注：相同脚注的氧化还原过程平均值（$n=3$）显著相关水平 $p<0.05$。

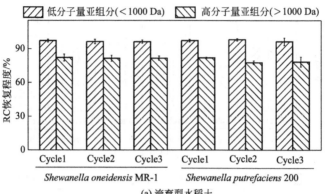
(a) 淹育型水稻土

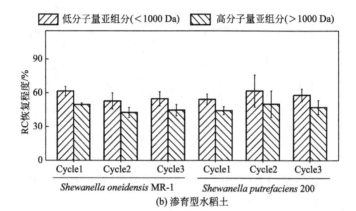
(b) 渗育型水稻土

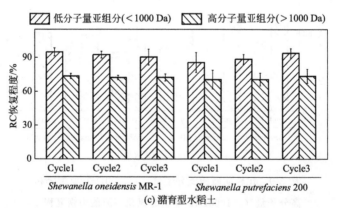
(c) 潴育型水稻土

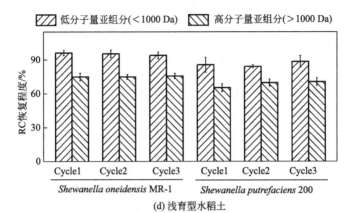

(d) 浅育型水稻土

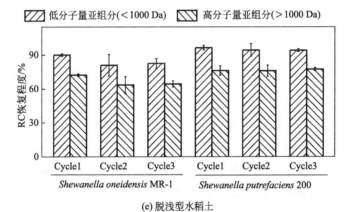

(e) 脱浅型水稻土

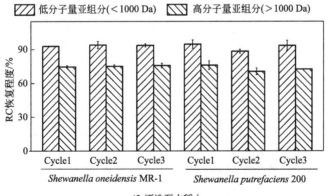

(f) 漂洗型水稻土

图 3-10

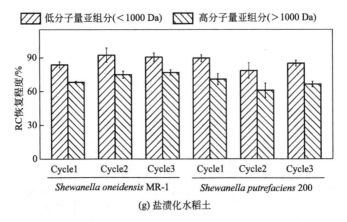

(g) 盐渍化水稻土

图3-10 比较富里酸低分子量（<1000Da）亚组分和高分子量（>1000Da）亚组分氧化还原能力恢复程度

注：相同脚注的氧化还原过程平均值（$n=3$）显著相关水平$p<0.05$。

根据假定的腐殖质的形状规则，腐殖质的平均分子量越大，腐殖质化合物分子的尺寸就越大（Tan，2014）。腐殖质较大的分子尺寸可能导致其空间结构具有更高的灵活性和可变性。考虑到腐殖质的平均分子大小与溶液中腐殖质的结构有关（Myneni et al.，1999；Piccolo et al.，2001），我们推测，与具有较小平均分子量的腐殖质相比，在微生物还原和O_2再氧化发生后，具有较大平均分子量的腐殖质在空间结构中更容易改变或折叠（见图3-11）。所以，腐殖质中的一些被微生物还原的氧化还原活性官能团可以更容易被腐殖质结构的其他部分所遮蔽。因此，这些被保护的氧化还原活性官能团可以避免遭受O_2的进一步再氧化。

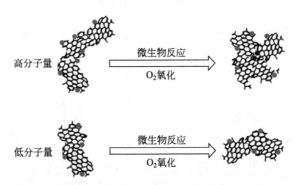

图3-11 微生物还原和O_2再氧化后腐殖质空间结构变化的示意

随着微生物还原和氧气再氧化过程，腐殖质空间结构发生了变化，腐殖质结构中不同官能团之间的相互作用可能会被触发，随后会干扰这种空间结构的变化（Maurice and Namjesnik-Dejanovic，1999；Hayes and Malcolm，1997），从而影响还原能力的恢复程度。然而，观察到还原能力的恢复程度与通过 FTIR 光谱测定的腐殖质的官能团（见图 3-7）无关，表明官能团不影响腐殖质从还原态恢复到氧化态。尽管我们得出结论，芳族体系如醌类是腐殖质中氧化还原活性官能团的主要贡献者，但非醌结构如含氮和含硫官能团的贡献亦不能忽略。如 Bauer 和 Kappler（2009）所述，如果不同的氧化还原活性官能团确实具有不同程度的抗 O_2 再氧化能力，腐殖质样品之间的各种氧化还原活性官能团也可能部分地导致还原能力恢复程度的变化。然而，还需要进一步的研究来评估氧化还原活性官能团的差异对通过 O_2 从还原态变为氧化态的腐殖质的影响。

3.3 微生物还原和氧气再氧化循环过程中腐殖质的电子循环能力

在缺氧/好氧交替过程中，无菌腐殖质对照组的氧化还原状态并没有发生系统变化 [见图 3-2（b）、(d)]。与第一次氧化还原循环中的一致，在第二次和第三次氧化还原循环的缺氧和好氧条件下的培育分别刺激了接种样品中腐殖质的微生物还原和 O_2 再氧化。腐殖质在第二次和第三次氧化还原循环中的微生物还原能力也与 C/H 和类腐殖质成分呈正相关，与 E_4/E_6 和 C5 呈负相关（见图 3-3）。这些发现表明，在连续的氧化还原循环中腐殖质中的芳香族一直是腐殖质在微生物还原过程中的主要电子受体。

与第一次氧化还原循环类似，第二次和第三次氧化还原循环中还原能力的恢复程度在腐殖质样品以及胡敏酸和富里酸之间显著不同（$p<0.05$）(见表 3-2)。这些恢复程度也与腐殖质的数均分子量和重均分子量呈显著负相关（见图 3-6）。此外，在第二次和第三次氧化还原循环中发生了具有低分子量的胡敏酸（或富里酸）亚组分的还原能力恢复程度高于高分子量组分的现象（见图 3-8~图 3-10）。这些结果表明，平均分子量在整个连续氧化还原循环中持续影响腐殖质通过 O_2 从还原态恢复到氧化态。预计在第一次氧化还原循环中由于微生物还原和 O_2 再氧化而发生了腐殖质空间结构中的平均分子量依赖性变化，这种变化在实验时间序列是不可逆转的，即使随着氧化还原循环周期的数量的增加也不会进一步增强。总体而言，研究结果表明，腐殖质的平均分子量是确定连续交替缺氧/好氧条件下腐殖质中涉及电子穿梭的氧化还原活性官能团的百分比的重要因素。

连续交替缺氧/好氧条件下腐殖质氧化还原循环的电子量取决于腐殖质的微生物还原能力，并与还原能力的恢复程度呈负相关。值得注意的是，腐殖质的微生物还原能力和还原能力恢复程度在氧化还原循环之间具有显著相关性，但没有显示出显著差异性［见图 3-2、图 3-12（a）和图 3-12（b）］。因此，参与腐殖质氧化还原循环的电子数量在氧化还原循环之间显示出显著的相关性，但没有显示出显著的差异性［见表 3-2 和图 3-12（c）］。这些结果表明，在微生物还原和 O_2 再氧化发生后，在连续的氧化还原循环之间，在还原态和氧化态之间切换时，腐殖质的氧化还原循环是完全可逆的并且对于腐殖质的未受保护的氧化还原活性官能团是可持续的。

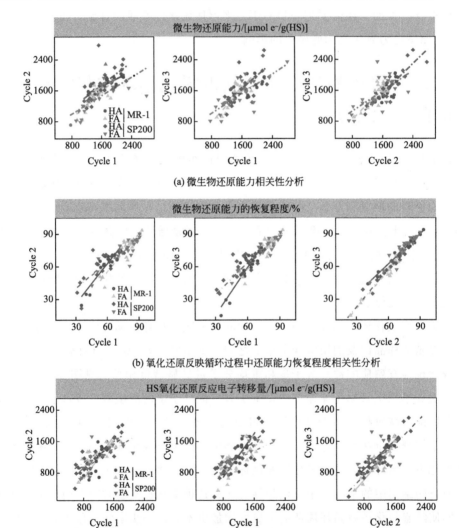

图 3-12　微生物还原能力及恢复程度、氧化还原反应电子转移量的相关性分析

注：图中线性显著性相关水平（$p<0.05$）。

3.4 胡敏酸与富里酸之间氧化还原循环能力的比较

基于胡敏酸与富里酸的比较，我们提出了两个结论。第一，胡敏酸的微生物还原能力显著大于富里酸［见图 3-13（a）］，这一观察主要归因于相比于富里酸，胡

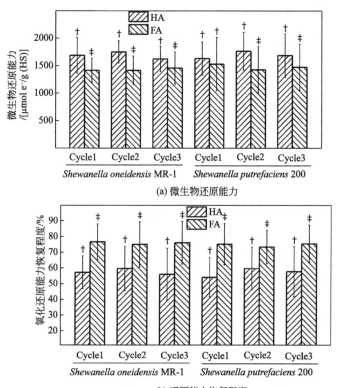

(a) 微生物还原能力

(b) 还原能力恢复程度

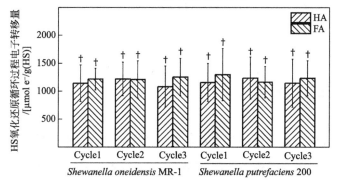

(c) 氧化还原循环过程电子转移量

图 3-13 比较胡敏酸和富里酸氧化还原循环中的微生物还原能力与恢复程度和氧化还原循环过程电子转移量

注：相同脚注表示无显著差异；不同脚注表示显著差异（$p<0.05$）。

敏酸中的含有较高的 C/H 值、HIX 和类腐殖质组分（C_4 和 C_6）和较低的 E_4/E_6 和 C5（见图 3-14）。第二，由于胡敏酸的数均分子量和重均分子量比富里酸高（见图 3-14），所以胡敏酸从还原态恢复为氧化态明显低于富里酸［见图 3-13（b）和表 3-2］。因此，上述两个结论导致涉及胡敏酸的氧化还原循环中的电子数与涉及交替缺氧/好氧条件下富里酸的氧化还原循环中的电子数没有显著差异［见图 3-13（c）和表 3-2］。这一发现表明，胡敏酸和富里酸的氧化还原循环在环境中起着同样重要的作用，如抑制短时缺氧系统中的 CH_4 产生。

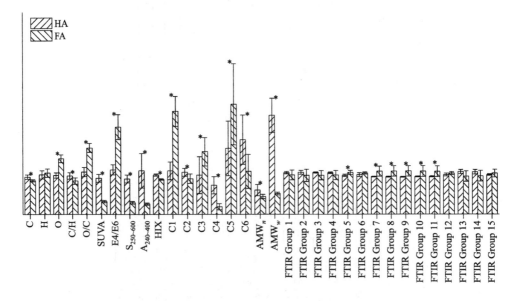

图 3-14　比较胡敏酸和富里酸化学结构参数

注：红外官能团组分 1~15 表示通过红外光谱分析的不同组分官能团（见表 3-1）；
每个参数标准化到相同的值。

3.5　环境意义

这项工作表明，在缺氧条件下腐殖质从铁还原菌接受电子的能力与其芳族结构呈正相关。在瞬时氧化条件下，O_2 刺激下的腐殖质从还原态到氧化态的恢复程度与腐殖质的 AMW 呈负相关。这些发现证实，源自水稻田土壤的腐殖质的化学结构是影响在交替缺氧/好氧过程中参与氧化还原循环的氧化还原活性官能团的百分比的关键因素，因而也影响了水稻田土中的 CH_4 形成。

作为两种重要的温室气体，CO_2 和 CH_4 在全球气候变化研究中得到广泛的重视（IPCC，2007）。分析和实验研究表明，有机物的化学结构不仅仅控制其在土壤中的分解和稳定（Knorr et al.，2005；Grandy and Neff，2008；Marschner et al.，2008）。因此，有一个新兴的观点强调，在评估土壤二氧化碳排放对全球变暖的影响因素时土壤有机质的化学结构可能是无关紧要的（Schmidt et al.，2011）。然而，从具有交替缺氧/好氧特征的土壤中 CH_4 形成的角度来看，我们建议将土壤有机质的化学结构信息纳入用于预测未来气温趋势的气候变化模型中，因为 CH_4 的形成可能与天然有机质的化学结构有关，如土壤中的腐殖质。

参考文献

Aeschbacher M，Sander M，Schwarzenbach R P，2010. Novel electrochemical approach to assess the redox properties of humic substances. Environmental Science & Technology，44（1）：87-93.

Aeschbacher M，Vergari D，Schwarzenbach R P，et al，2011. Electrochemical analysis of proton and electron transfer equilibria of the reducible moieties in humic acids. Environmental Science & Technology，45（19）：8385-8394.

Bauer I，Kappler A，2009. Rates and extent of reduction of Fe(Ⅲ) compounds and O_2 by humic substances. Environmental Science & Technology，43（13）：4902-4908.

Benz M，Schink B，Brune A，1998. Humic acid reduction by propionibacterium freudenreichii and other fermenting bacteria. Applied and Environmental Microbiology，64（11）：4507-4512.

Blodau C，Deppe M，2012. Humic acid addition lowers methane release in peats of the Mer Bleue bog，Canada. Soil Biology & Biochemistry，52：96-98.

Bridgham S D，Cadillo-Quiroz H，Keller J K，et al，2013. Methane emissions from wetlands：Biogeochemical，microbial，and modeling perspectives from local to global scales. Global Change Biology，19（5）：1325-1346.

Cervantes F J，de Bok F A M D，Tuan D D，et al，2002. Reduction of humic substances by halorespiring，sulphate-reducing and methanogenic microorganisms. Environmental Microbiology，4（1）：51-57.

Chen J，Gu B，Royer R A，et al，2003. The roles of natural organic matter in chemical and microbial reduction of ferric iron. Science of the Total Environment，307（1-3）：167-178.

Dunnivant F M，Schwarzenbach R P，Macalady D L，1992. Reduction of substituted nitrobenzenes in aqueous solutions containing natural organic matter. Environmental Science & Technology，26（11）：2133-2141.

Duval J F L，Wilkinson，K J，Van Leeuwen H P，et al，2005. Humic substances are soft and permeable：evidence from their electrophoretic mobilities. Environmental Science & Technology, 39 (17): 6435-6445.

Einsiedl F, Mayer B, Thorsten S, 2008. Evidence for incorporation of H_2S in groundwater fulvic acids from stable isotope ratios and sulfur K-edge X-ray absorption near edge structure spectroscopy. Environmental Science & Technology, 42 (7): 2439-2444.

Engebretson R R, Von Wandruszka R, 1994. Micro-Organization in dissolved humic acids. Environmental Science & Technology, 28 (11): 1934-1941.

Gu B, Chen J, 2003. Enhanced microbial reduction of Cr(Ⅵ) and U(Ⅵ) by different natural organic matter fractions. Geochimica et Cosmochimica Acta, 67 (19): 3575-3582.

Grandy A S, Neff J C, 2008. Molecular C dynamics downstream: the biochemical decomposition sequence and its impact on soil organic matter structure and function. Science of the Total Environment, 404 (2-3): 297-307.

Hayes M H B, Malcolm R L, 1997. Considerations of compositions and aspects of structures of humic substances. Soil Science Society of America Journal, 3-39.

Herzsprung P, von Tümpling W, Hertkorn N, et al, 2012. Variations of DOM quality in inflows of a drinking water reservoir: Linking of van krevelen diagrams with EEMF spectra by rank correlation. Environmental Science & Technology, 46 (10): 5511-5518.

IPCC, 2007. Climate Change 2007: The physical scientific basis. The Fourth Assessment Report of Working Group. Cambridge: Cambridge UnivPress.

Kappler A, Haderlein S B, 2003. Natural organic matter as reductant for chlorinated aliphatic pollutants. Environmental Science & Technology, 37 (12): 2714-2719.

Keller J K, Weisenhorn P B, Megonigal J P, 2009. Humic acids as electron acceptors in wetland decomposition. Soil Biology & Biochemistry, 41 (7): 1518-1522.

Kleber M, Johnson M G, 2010. Advances in understanding the molecular structure of soil organic matter: Implications for interactions in the environment. Advances in Agronomy, 106: 77-142.

Klüpfel L, Piepenbrock A, Kappler A, et al, 2014. Humic substances as fully regenerable electron acceptors in recurrently anoxic environments. Nature Geoscience, 7 (3): 195-200.

Knorr M, Frey S D, Curtis P S, 2005. Nitrogen additions and litter decomposition: a meta-analysis. Ecology, 86 (12): 3252-3257.

Lovley D R, Bluntharris E L, Ejp P, et al, 1996. Humic substances as electron acceptors for microbial respiration. Nature, 382 (6590): 445-448.

Marschner B, Brodowski S, Dreves A, et al, 2008. How relevant is recalcitrance for the stabilization of organic matter in soils. Journal of Plant Nutrition and Soil Science, 171 (1): 91-110.

Maurice P A, Namjesnik-Dejanovic K, 1999. Aggregate structures of sorbed humic substances observed in aqueous solution. Environmental Science & Technology, 33 (9): 1538-1541.

Myneni S C B, Brown J T, Martinez G A, et al, 1999. Imaging of humic substance macromolecular structures in water and soils. Science, 286 (5443): 1335-1337.

Miller K E, Lai C T, Friedman E S, et al, 2015. Methane suppression by iron and humic acids in soils of the arctic coastal plain. Soil Biology & Biochemistry, 83: 176-183.

Nakayasu K, Fukushima M, Sasaki K, 1999. Comparative studies of the reduction behavior of chromium- (Ⅵ) by humic substances and their precursors. Environmental Toxicology and Chemistry, 18 (6): 1085-1090.

Ohno T, 2002. Fluorescence inner-filtering correction for determining the humification index of dissolved organic matter. Environmental Science and Technology, 36 (4): 742-746.

Piccolo A, Conte P, Cozzolino A, 2001. Chromatographic and spectrophotometric properties of dissolved humic substances compared with macromolecular polymers. Soil Science, 166 (3): 174-185.

Ratasuk N, Nanny M A, 2007. Characterization and quantification of reversible redox sites in humic substances. Environmental Science & Technology, 41 (22): 7844-7850.

Schmidt M W I, Torn M S, Abiven S, et al, 2011. Persistence of soil organic matter as an ecosystem property. Nature, 478 (7367): 49-56.

Schwarzenbach R P, Gschwend P M, Imboden D M, 2005. Environmental organic chemistry. New York: John Wiley & Sons.

Scott D T, Mcknight D M, Blunt-Harris E L, et al, 1998. Quinone moieties act as electron acceptors in the reduction of humic substances by humics-reducing microorganisms. Environmental Science & Technology, 32 (19): 2984-2989.

Stevenson F J, 1994. Humus Chemistry. New York: John Wiley & Sons.

Struyk Z, Sposito G, 2001. Redox properties of standard humic acids. Geoderma, 102 (3-4): 329-346.

Tan W, Xi B, Wang G, et al, 2017. Increased electron-accepting and decreased electron-donating capacities of soil humic substances in response to increasing temperature. Environmental Science & Technology, 51 (6): 3176-3186.

Tan K H, 2014. Humic matter in soil and the environment: Principles and controversies (Second ed.). Boca Raton: CRC Press.

Thieme J, Mcnult I, Vogt S, et al, 2007. X-Ray spectromicroscopy-A tool for environmental sciences. Environmental Science & Technology, 41 (20): 6885-6889.

Tokida T, Adachi M, Cheng W, et al, 2011. Methane and soil CO_2 production from current-season photosynthates in a rice paddy exposed to elevated CO_2 concentration and soil temperature. Global Change Biology, 17 (11): 3327-3337.

Wittbrodt P R, Palmer C D, 1997. Reduction of Cr(Ⅵ) by soil humic acids. European Journal of Soil Science, 48 (1): 151-162.

Van der Zee F R, Cervantes F J, 2009. Impact and application of electron shuttles on the redox (bio) transformation of contaminants: A review. Biotechnology Advances, 27 (3): 256-277.

Von Wandruszka R, Engebretson R R, Yates L M, 1999. Humic acid pseudomicelles in dilute aqueous solution: Fluorescence and surface tension measurements. Understanding Humic Substances: 79-85.

Ye R, Doane T A, Morris J, et al, 2015. The effect of rice straw on the priming of soil organic matter and methane production in peat soils. Soil Biology & Biochemistry, 81: 98-107.

Zhang G, Yu H, Fan X, et al, 2015. Effect of rice straw application on stable carbon isotopes, methanogenic pathway, and fraction of CH_4 oxidized in a continuously flooded rice field in winter season. Soil Biology & Biochemistry, 84: 75-82.

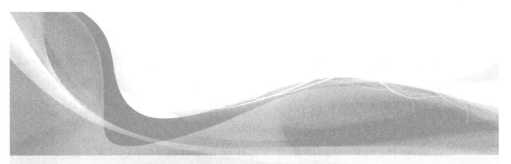

第4章 土壤腐殖质电子转移能力对增温的响应

土壤腐殖质是指高等植物、微生物与动物遗体通过微生物降解所形成的天然有机大分子的不均匀混合物（Stevenson，1994）。由于缺少腐殖化进程的证据，新兴的土壤连续介质模型驳斥了"HS"的概念，而将土壤有机质看作一种跨越从氧化碳到完整植物的连续统一体（Lehmann and Kleber，2015）。但仍不可否认腐殖质是土壤提取工艺的一种操作上的定义。同时，由于腐殖质的氧化还原性质在环境相关的过程中具有重要意义，其氧化还原性质的研究在众多的环境基质中就显得尤为重要（Klüpfel et al.，2014；Piepenbrock et al.，2014；Piepenbrock and Kappler，2013）。处于还原态、溶解态和颗粒态的腐殖质可以直接从微生物上得到电子（Lovley et al.，1996；Roden et al.，2010），像铁还原菌（Lovley et al.，1996）、硫酸盐还原菌（Cervantes et al.，2002）和发酵菌（Benz et al.，1998）等。还原态的腐殖质可以将电子供给不易得电子的氧化铁和氢氧化物（Lovley et al.，1996），以及各种有机和无机污染物（Borch et al.，2010；Alberts et al.，1974），包括氯化化合物（Kappler and Haderlein，2003）、硝基苯（Dunnivant et al.，1992；Van der Zee and Cervantes，2009）、U(Ⅵ)（Gu and Chen，2003）和Cr(Ⅵ)（Nakayasu et al.，1999；Wittbrodt and Palmer，1997）。因此，腐殖质的电子转移能力可以显著影响氧化还原活性污染物的生物地球化学氧化还原过程。

土壤腐殖质的电子转移能力（ETC）主要归因于其结构内富含各种氧化还原活性官能团这个固有的物理化学性质（Scott et al.，1998；Aeschbacher et al.，2012；Struyk and Sposito，2001；Chen et al.，2003；Einsiedl et al.，2008）。电子自旋共振光谱的直接证据显示，醌基是微生物还原过程中接受电子的官能团（Scott et al.，1998）。通过对模型醌类和腐殖质的光谱性质比较，表明了醌基涉及了腐殖质的还原过程（Ratasuk and Nanny，2007）。傅里叶变换红外光谱，核磁共振和热解气相色谱-质谱法也指出醌基是腐殖质中重要的氧化还原功能基团

(Hernández-Montoya et al.，2012；Aeschbacher et al.，2010）。此外，由于基于给定腐殖质质量与滴定的苯酚含量所贡献的电子的强线性相关性（Aeschbacher et al.，2012；Walpen et al.，2016）和基于苯酚-对苯二酚混合物与腐殖质氧化还原滴定曲线的氧化还原电位的相似 pH 依赖性，认为苯酚官能团也是腐殖质中的主要电子供体功能基团（Helburn and Maccarthy，1994）。

天然有机物（NOM）和腐殖质的固有物理化学性质在很大程度上取决于它们在土壤中的氧化转化和降解（Stevenson，1994；Kleber and Johnson，2010）。自然温度作为一项状态因子，可用来预测土壤微环境条件、微生物群落结构、活跃度以及植物凋落物质量的变化（Swift et al.，1979；Conant et al.，2011；Han et al.，2011）。这种变化可能对氧化转化和降解以及最终对土壤中的腐殖质的电子转移能力产生重大影响。因此，土壤中腐殖质的归宿和功能本质上与自然温度有关。确定自然温度改变对土壤中腐殖质电子转移能力的影响机制，有利于更好地了解自然温度对土壤腐殖质氧化还原功能相关的生物地球化学作用过程。

中国黄土高原的黄土-古土壤序列包含有层状黄土和埋藏的古土壤单元，被认为分别是经历了主要偏心节奏的冰川和间冰期气候的时期（Kukla et al.，1990；An et al.，1990）。然而，重建黄土-古土壤序列在地质年代中的古地温的数据却难以获取（Lu et al.，2007）。采用"空间替代时间"的方法，例如跨越纬度或海拔梯度的土壤采样，具有自然气候梯度的固有优势（Parmesan and Yohe，2003），可以预测温度升高对土壤腐殖质电子转移能力的影响。取自不同维度梯度的土壤，可以消除降水量不同引起的干扰，但会导致土壤的差异性增大，取自不同海拔梯度的土壤虽然会消除土壤母质的差异性，但不能保持该地的年平均降水量始终保持一致，每种采样方式的内在优势可以抵消其他采样方式的缺点。

在本书中，从中国洛川的黄土-古土壤序列中采集了两类土壤：一类是沿着中国西北至西南的 400 mm 平均年降水等值线的纬度梯度的土壤；另一类是中国东陵山高原海拔梯度的土壤。本书旨在：a.确定影响腐殖质的电子接受能力（EAC）和电子供给能力（EDC）的化学结构；b.探讨土壤物理化学性质和植物凋落与土壤腐殖质电子转移能力的关系；c.研究土壤和植物凋落物的物理化学性质与温度变化的关系；d.评估土壤腐殖质电子转移能力与温度变化的关系。本书结论可以提高我们在温度升高背景下认识腐殖质和天然有机质的氧化还原性质在土壤生物地球化学过程中的重要作用。

4.1　土壤腐殖质电子转移能力与温度的关系

选择 27 个和 18 个采样点分别沿中国西北至西南的 400mm 年平均降水等值线

(29.06~53.29°N, 90.39~122.15°E; 纬度梯度, 年平均气温在-5.5℃和8.9℃) 和东岭山 (39.92~40.03°N, 115.45~115.57°E; 海拔梯度, 平均气温在-0.7~10.5℃之间) 采样。纬度和海拔梯度的采样点的表面土壤温度不受特定地质条件 (例如地热) 的影响, 而主要受空气温度控制。纬度梯度上的27个采样点和高程梯度上的18个采样点的地理位置和年平均气温如表4-1、表4-2所列。在纬度和海拔梯度的每个地点, 从地块 (10m×10m) 中随机选出10个样本 (包括植物凋落物和最深10cm的土壤) 并混合。所有采样点远离人类栖息地, 以尽量减少人类活动的影响。所有的土壤和植物凋落物样品从采样到分析, 期间保持在-20℃环境中保存。PARAFAC分析确定了6个组分 [图4-1 (a)], 其独立地覆盖所有土壤组。

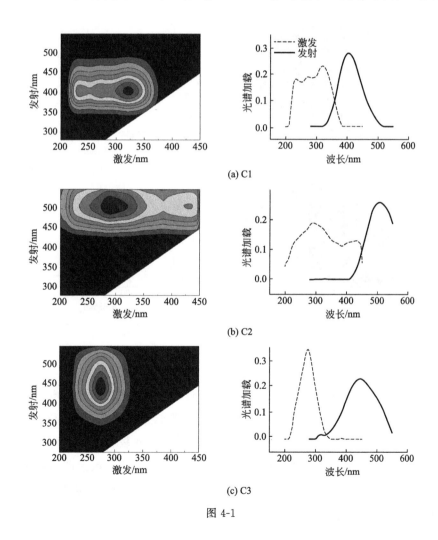

图 4-1

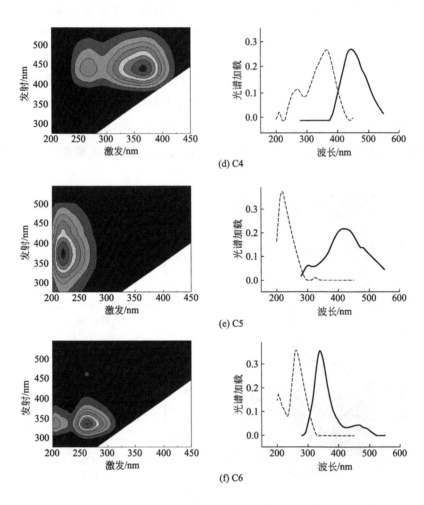

图 4-1 PARAFAC 分析确定的六种组分验证模型的激发和发射负载（见书后彩图 7）

在黄土-古土壤（冰川-间冰期）序列中，胡敏酸和富里酸两者的电子转移能力显示，相对于黄土，古土壤中具有更高的电子接受能力和较低的电子供给能力的周期性波动特征（图 4-2）。尽管缺乏有效的方法来定量比较古土壤期与黄土期之间的古温差，估计在古土壤形成期间，平均空气温度高于黄土积累期间的平均气温（Peterse et al.，2011）。因此，结果表明，高温有利于增加土壤腐殖质的电子接受能力并降低电子供给能力。

表 4-1 不同纬度土壤样品的基本信息

采样序号	纬度/(°)	经度/(°)	海拔高度/m	年平均气温/℃	主要植被类型	主要生物类型	主要土壤类型
1	53.29	122.15	603	-5.5	寒带温带针叶林	落叶松	棕色针叶林土壤
2	50.46	121.31	718	-5.3	寒带温带针叶林	白桦	棕色针叶林土壤
3	49.94	121.43	829	-5.5	寒带温带针叶林	东北白桦	棕色针叶林土壤
4	49.33	120.97	634	-2.9	寒带温带针叶林	落叶松	棕色针叶林土壤
5	47.10	119.89	1240	-2.7	寒带温带针叶林	东北白桦	棕色针叶林土壤
6	46.05	121.79	366	4.1	温带草原	针茅、羊草、线叶菊属	黑钙土
7	44.61	120.97	332	2.8	温带草原	针茅、羊草、线叶菊属	黑钙土
8	43.44	120.08	928	2.2	温带草原	针茅、羊草	栗土
9	42.18	116.47	1245	2.4	温带草原	针茅、羊草	栗土
10	40.45	113.19	1236	4.7	温带草原	针茅、羊草	栗土
11	39.87	111.18	1236	7.5	温带草原	针茅、羊草	棕色钙土
12	39.33	111.19	912	8.8	温带草原	狼尾草	棕色钙土
13	38.84	110.44	1131	8.9	半荒漠草原	针茅、女蒿、亚菊	灰钙土
14	38.04	109.24	1131	8.5	半荒漠草原	针茅、女蒿、亚菊	灰钙土
15	37.74	108.91	1394	7.8	半荒漠草原	针茅、女蒿、亚菊	灰钙土
16	36.02	105.88	1982	5.3	温带草原	针茅、女蒿	栗钙土
17	35.78	104.05	2361	6.6	温带针叶林	松树	棕色针叶林甸草地土壤
18	36.69	101.3	3233	1.2	亚高山草甸	针茅、女蒿	亚高山草甸土壤
19	37.02	100.8	2725	1.5	亚高山草原	针茅、女蒿	亚高山草原土壤
20	35.55	102.03	2467	5.2	亚高山草原	针茅、女蒿	亚高山草原土壤
21	35.27	100.64	3258	-0.5	亚高山草原	针茅、女蒿	亚高山草原土壤
22	34.16	95.90	4727	-5.3	高寒草甸	蒿草	高寒草原土壤
23	33.77	95.66	4360	-1.2	高寒草甸	蒿草	高寒草原土壤
24	32.09	92.27	4731	-2.2	高寒草甸	蒿草	高寒草原土壤
25	31.41	91.96	4519	-3.2	高寒草甸	蒿草	亚高山草原土壤
26	29.33	88.98	3865	6.3	高寒草甸	针茅、羊茅、薹草	高寒草原土壤
27	29.06	90.39	4492	3.0	高寒草甸	针茅、羊茅、薹草	高寒草原土壤

表 4-2 不同海拔梯度土壤样品的基本信息

采样序号	纬度/(°)	经度/(°)	海拔高度/m	年平均气温/℃	主要植被类型	主要生物类型	主要土壤类型
1	39.926	115.569	400	10.5	灌木草原	狗尾草	褐土
2	39.927	115.561	600	9.0	灌木草原	狗尾草、藜	褐土
3	39.931	115.552	730	8.0	灌木草原	狗尾草、藜	褐土
4	39.936	115.545	800	7.5	灌木草原	狗尾草、车前子、藜	褐土
5	39.940	115.541	900	6.8	灌木草原	狗尾草、车前子	褐土
6	39.945	115.540	1020	5.9	灌木草原	狗尾草、车前子	棕壤
7	39.961	115.525	1120	5.2	灌木草原	狗尾草、车前子	棕壤
8	39.966	115.523	1280	4.6	灌木草原	狗尾草、车前子	棕壤
9	39.971	115.520	1380	4.2	灌木草原	狗尾草、车前子	棕壤
10	39.995	115.504	1480	3.8	灌木草原	车前子、地榆	棕壤
11	40.001	115.505	1580	3.4	灌木草原	地榆、瓣蕊唐松草	棕壤
12	40.004	115.509	1680	3.0	灌木草原	地榆、瓣蕊唐松草	棕壤
13	40.012	115.462	1800	2.3	亚高山草甸	地榆、瓣蕊唐松草、直穗鹅观草	草甸土
14	40.018	115.460	1880	1.8	亚高山草甸	地榆、瓣蕊唐松草、直穗鹅观草	草甸土
15	40.022	115.460	2000	1.1	亚高山草甸	车前子、地榆、瓣蕊唐松草	草甸土
16	40.026	115.463	2100	0.5	亚高山草甸	车前子、地榆、瓣蕊唐松草	草甸土
17	40.030	115.457	2200	−0.1	亚高山草甸	地榆、瓣蕊唐松草	草甸土
18	40.029	115.450	2300	−0.7	亚高山草甸	地榆	草甸土

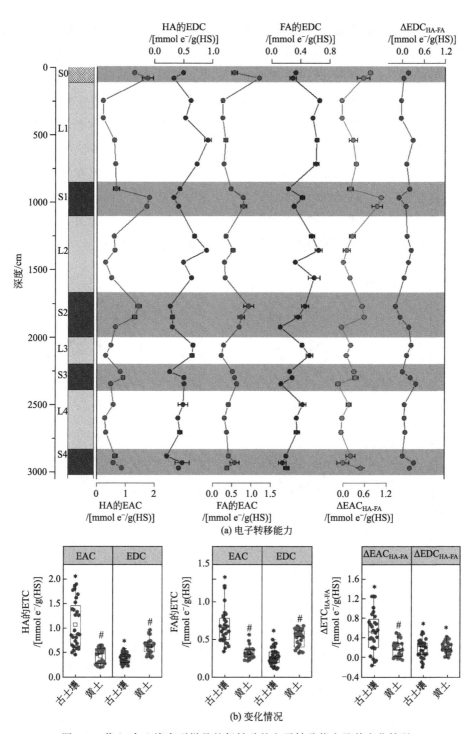

图 4-2 黄土-古土壤序列样品的胡敏酸的电子转移能力及其变化情况

横跨纬度和海拔梯度的土壤也表明，土壤腐殖质的电子转移能力与年平均气温显著相关。胡敏酸和富里酸的电子接受能力和电子供给能力分别随着温度的升高而逐渐增加和降低［图4-3（a）、(d) 和图4-4（a）、(d)］。当纬度和海拔数据集合成一个数据集时，土壤腐殖质的电子转移能力与平均年温度之间的关系仍然很大［图4-5（a）～(d)］，表明温度对土壤腐殖质的电子转移能力的影响是普遍的。这些基于黄土-古生物序列之间的土壤的发现进一步证实了上述结论。在所有土壤组中，在大多数样品［图4-2、图4-3（e）～(f)、图4-4（e）～(f) 和图4-5（e）～(f)］中，胡敏酸具有比富里酸更大的电子接受能力和电子供给能力。尽管随着温度升高，胡敏酸与富里酸的电子接受能力趋势一致，但胡敏酸和富里酸之间的电子接受能力差异与年平均气温呈正相关［图4-2（b）、图4-3（e）、图4-4（e）和图4-5（e）］。相比于土壤富里酸，土壤胡敏酸的电子接受能力更依赖于温度变化。

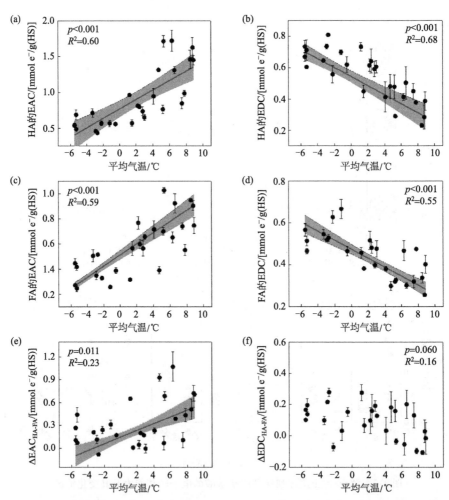

图4-3　不同维度梯度的土壤腐殖质的电子转移能力及其与年均温度的响应关系

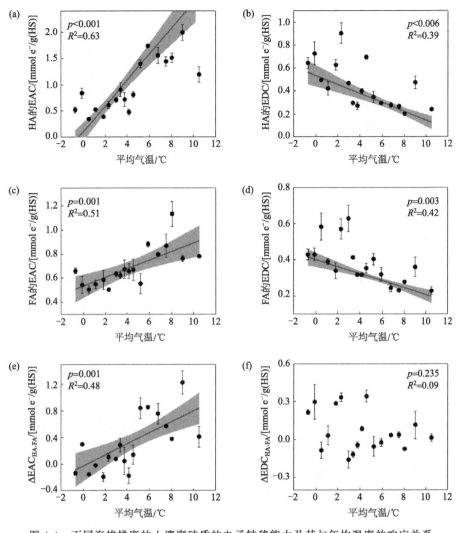

图 4-4　不同海拔梯度的土壤腐殖质的电子转移能力及其与年均温度的响应关系

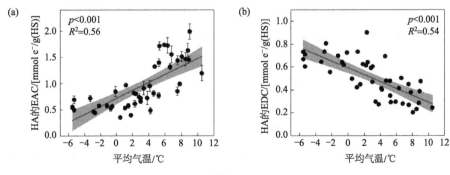

图 4-5

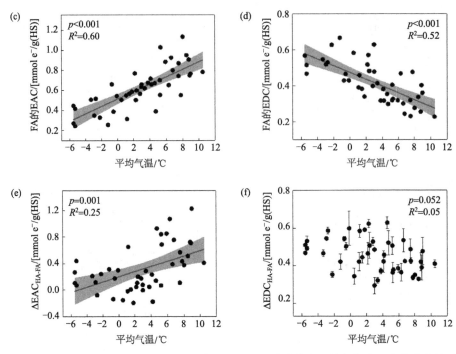

图 4-5　从纬度到海拔梯度的整个土壤腐殖质的电子转移能力及其与年均温度的响应关系

4.2　土壤腐殖质电子转移能力与其化学结构的关系

我们采用了广泛的参数（见表 4-3）反映土壤腐殖质的化学结构，以阐明土壤腐殖质的电子转移能力与其固有的物理化学性质之间的关系。土壤腐殖质的电子接受能力与芳基 C、C/H 值、C2 和 HIX 呈正相关，与烷基 C、H、E_4/E_6 和 C6 呈负相关（图 4-6），而土壤腐殖质的电子供给能力与滴定的苯酚和木质素衍生的丁香醛、丁香酸和香草醛呈正相关（图 4-6）。土壤腐殖质的电子转移能力与其物理化学性质之间的关系在每个独立土壤组合和整体土壤中都是通用的，这些土壤集合了沿黄土-古生物序列的土壤、纬度和海拔梯度的土壤、胡敏酸和富里酸的独立数据集，以及胡敏酸和富里酸的集成数据集都适用（图 4-6）。结果表明，上述物理化学性质与土壤腐殖质的电子转移能力的因果关系与土壤来源和腐殖质组分无关。采用逐步多元回归方法来识别导致电子转移能力变化的土壤腐殖质的化学结构，芳基 C、C/H 值、滴定苯酚、丁香醛、香草醛和 E_4/E_6 始终是土壤腐殖质电子转移能力的最佳预测因子（图 4-7），分别解释了不同土壤组和腐殖质组分中电子接受能力和电子功能能力变化的 54%～74% 和 40%～56%。

表 4-3　土壤、植物凋落物及腐殖质物理化学性质的指标摘要

变量	说明	单位
EAC	腐殖质的电子接受能力	mmole$^-$/g(HS)
EDC	腐殖质的电子供给能力	mmole$^-$/g(HS)
Alkyl C	烷基 C,通过 13NMR 光谱估计 HS 的分子结构	%
O-alkyl C	邻烷基 C,通过 13NMR 光谱估计 HS 的分子结构	%
Aryl C	芳基 C,通过 13NMR 光谱估计 HS 的分子结构	%
Carboxylic C	羧酸 C,通过 13NMR 光谱估计 HS 的分子结构	%
HS titrated phenol	滴定苯酚,HS 中的总酚	mmole$^-$/g(HS)
HS p-coumaric acid	对香豆酸,HS 的木质素衍生酚	mg/g(HS)
HS ferulic acid	阿魏酸,HS 的木质素衍生酚	mg/g(HS)
HS syringaldehyde	丁香醛,HS 的木质素衍生酚	mg/g(HS)
HS acetosyringone	乙酰丁香酮,HS 的木质素衍生酚	mg/g(HS)
HS syringic acid	丁香酸,HS 的木质素衍生酚	mg/g(HS)
HS vanillin	香兰素,HS 的木质素衍生酚	mg/g(HS)
HS acetovanillone	乙酰香草醛,HS 的木质素衍生酚	mg/g(HS)
HS vanillic acid	香草酸,HS 的木质素衍生酚	mg/g(HS)
HS element C	HS 的元素组成	g/kg(HS)
HS element H	HS 的元素组成	g/kg(HS)
HS element O	HS 的元素组成	g/kg(HS)
HS element N	HS 的元素组成	g/kg(HS)
HS element S	HS 的元素组成	g/kg(HS)
HS element C/H	HS 的元素比例	—
HS element O/C	HS 的元素比例	—
HS element(N+S)/C	HS 的元素比例	—
SUVA$_{254}$	254nm 的比紫外线吸收率	L/(m·mg)
E_4/E_6	465nm 和 665nm 的紫外线可见吸收率的比	—
$A_{240\sim400}$	240~400nm 的紫外光谱面积	—
$S_{275\sim295}$	275~295nm 的紫外光谱斜率	—
$S_{350\sim400}$	350~400nm 的紫外光谱斜率	—
SR	$S_{275\sim295}/S_{350\sim400}$ 比例	—

续表

变量	说明	单位
C1	类腐殖质荧光组分	%
C2	类腐殖质荧光组分	%
C3	类腐殖质荧光组分	%
C4	类腐殖质荧光组分	%
C5	微生物衍生荧光组分	%
C6	类蛋白质荧光组分	%
HIX	腐殖化指数	—
HS content	土壤中 HS 含量	g/kg(土壤)
Soil C/N	土壤中元素比	—
Soil clay content	土壤中黏土含量	%
Soil microbial biomass	土壤微生物量,土壤的酶活性	g(C)/[kg(土壤)]
Soil catalase	土壤过氧化氢酶,土壤的酶活性	mL(20mmol/L $KMnO_4$)/[h·g(土壤)]
Soil lignin peroxidase	土壤木质素过氧化物酶,土壤的酶活性	μmol/[min·g(土壤)]
Soil manganese peroxidase	土壤锰过氧化物酶,土壤的酶活性	μmol/[min·g(土壤)]
Soil laccase	土壤漆酶,土壤的酶活性	μmol/[min·g(土壤)]
Soil β-glucosidase	土壤 β-葡萄糖苷酶,土壤的酶活性	μmol/[min·g(土壤)]
Soil dehydrogenase	土壤脱氢酶,土壤的酶活性	μmol/[min·g(土壤)]
Soil urease	土壤脲酶,土壤的酶活性	μmol/[min·g(土壤)]
Soil alkaline phosphatase	土壤碱性磷酸酶,土壤的酶活性	μmol/[min·g(土壤)]
Soil N	土壤的元素组成	g/kg(土壤)
Soil P	土壤的元素组成	g/kg(土壤)
Soil K	土壤的元素组成	g/kg(土壤)
Soil Ca	土壤的元素组成	g/kg(土壤)
Soil Mg	土壤的元素组成	g/kg(土壤)
Soil S	土壤的元素组成	g/kg(土壤)
Soil Fe	土壤的元素组成	g/kg(土壤)
Soil Mn	土壤的元素组成	g/kg(土壤)
Soil Cu	土壤的元素组成	g/kg(土壤)

续表

变量	说明	单位
Soil Zn	土壤的元素组成	g/kg(土壤)
Soil Mo	土壤的元素组成	g/kg(土壤)
Litter p-coumaric acid	对香豆酸,植物凋落物的木质素衍生酚	mg/g(C)
Litter ferulic acid	阿魏酸,植物凋落物的木质素衍生酚	mg/g(C)
Litter syringaldehyde	丁香醛,植物凋落物的木质素衍生酚	mg/g(C)
Litter acetosyringone	乙酰丁香酮,植物凋落物的木质素衍生酚	mg/g(C)
Litter syringic acid	丁香酸,植物凋落物的木质素衍生酚	mg/g(C)
Litter vanillin	香兰素,植物凋落物的木质素衍生酚	mg/g(C)
Litter acetovanillone	乙酰香草醛,植物凋落物的木质素衍生酚	mg/g(C)
Litter vanillic acid	香草酸,植物凋落物的木质素衍生酚	mg/g(C)
Litter N	植物凋落物的元素组成	g/kg(凋落物)
Litter P	植物凋落物的元素组成	g/kg(凋落物)
Litter K	植物凋落物的元素组成	g/kg(凋落物)
Litter Ca	植物凋落物的元素组成	g/kg(凋落物)
Litter Mg	植物凋落物的元素组成	g/kg(凋落物)
Litter S	植物凋落物的元素组成	g/kg(凋落物)
Litter Fe	植物凋落物的元素组成	g/kg(凋落物)
Litter Mn	植物凋落物的元素组成	g/kg(凋落物)

高 C/H 值或高 HIX 通常表明在天然有机质和腐殖质中的芳香环高度聚合（Ohno，2002；Stevenson，1994）。荧光和傅里叶变换离子回旋共振质谱（FT-ICR-MS）数据的结果表明，类腐殖质荧光通常与芳香族结构共同变化（Herzsprung et al.，2012）。低 E_4/E_6 主要是由于具有芳族碳-碳双键的官能团的吸收（Kleber and Johnson，2010）。除了滴定苯酚，丁香醛、丁香酸和香草醛也是重要的单环酚化合物（Thevenot et al.，2010）。上述信息表明 C/H 值、HIX、C2 和 E_4/E_6 反映了腐殖质中的醌类结构，滴定苯酚、丁香醛、丁香酸和香草醛反映了腐殖质中的酚类结构。因此，我们的研究结果表明，芳香族体系在腐殖质中起氧化还原活性官能团的作用，与醌和酚分别是腐殖质中主要的电子接受和电子供给官能团的观点一致（Lovley et al.，1996；Scott et al.，1998；Aeschbacher et al.，2012；Walpen et al.，2016）。此外，考虑到微生物死亡后产生的物质可能会增加土壤中

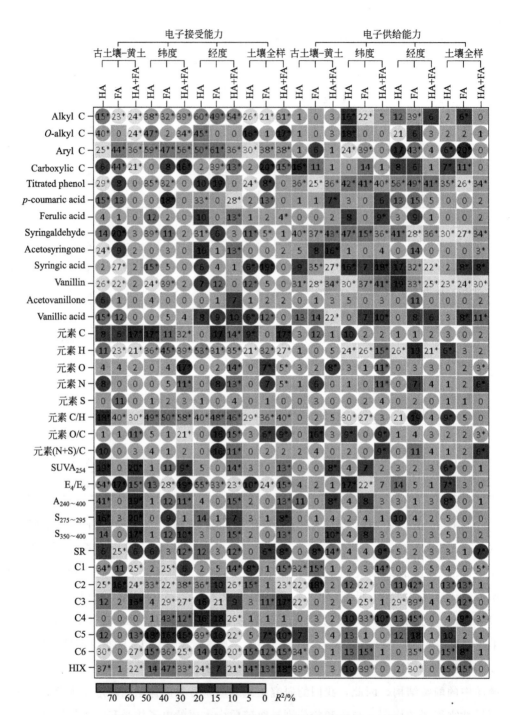

图 4-6 土壤腐殖质的电子供给能力和电子接受能力与其物理化学结构的相关系数（见书后彩图 8）

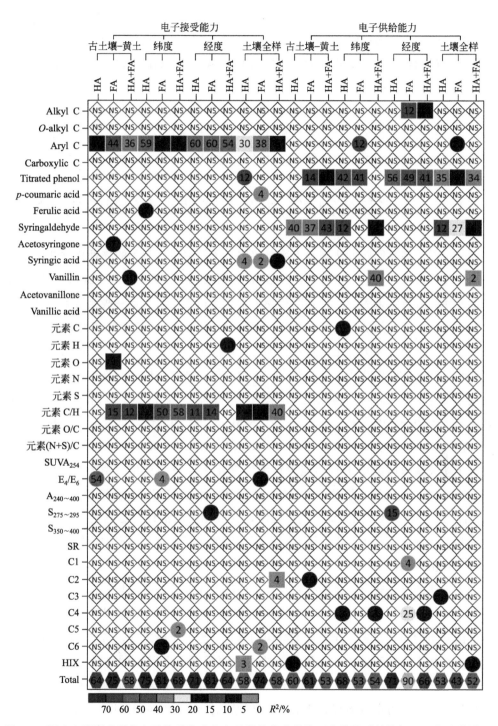

图 4-7 影响土壤腐殖质的电子接受能力和电子供给能力的物理化学性质的指标（见书后彩图 9）

具有更长周转时间的天然有机质和腐殖质（Kiem and Koögel-Knabner，2003；Kindler et al.，2006；Liang and Balser，2008；Simpson et al.，2007），衍生自微生物死亡后的物质，如氧化还原辅因子［例如铁-硫簇（Imsande，1999）、核黄素（Marsili et al.，2008）、黄素单核苷酸（Okamoto et al.，2013）和钼蝶呤鸟嘌呤二核苷酸（Palmer et al.，2005）］、细胞色素（White et al.，2013）和黑色素（Turick et al.，2002），也可能与腐殖质中可能的氧化还原活性化合物相关。然而，量化这些功能组对土壤中腐殖质电子转移能力的贡献有待进一步研究。

4.3　土壤和植物凋落物对土壤腐殖质化学结构的影响

采用逐步多元回归方法确定土壤腐殖质电子转移能力相关的物理化学性质与表征跨维度或海拔梯度的土壤和植物枯枝落叶特征的指标（表4-1）之间的关系。逐步多元回归分析结果表明，植物凋落物中木质素衍生的酚类是一些土壤组的腐殖质中木质素衍生的酚类的正向预测因子（图4-8～图4-11），这意味着植物凋落物中的一些木质素衍生酚类是土壤腐殖质中木质素衍生酚的重要来源。土壤腐殖质的形成和转化通常由微生物驱动。尽管土壤腐殖质的物理化学性质与跨纬度和海拔梯度土壤中微生物量之间缺乏强相关性（图4-8～图4-11），但逐步多元回归分析结果表明，微生物分泌的过氧化氢酶、LiP、MnP和Lac的活性，负责木质素的氧化转化（Kirk and Farrell，1987；Tien and Kirk，1983；Leonowicz et al.，2001），这解释了涉及腐殖质的电子转移能力相关物理化学性质的变化（图4-8～图4-11），例如芳基C、C/H值、组分C2、HIX的增加，以及烷基C、滴定苯酚、丁香醛、H、E_4/E_6和组分C6的降低。结果表明，这些相关酶活性是影响腐殖质电子转移能力相关物理化学性质的重要因素，对醌基团的含量产生积极影响，但对苯酚基团含量表现出负面影响。

腐殖质的氧化转化被认为伴随着高浓度酚基团初始分级产物氧化为醌基团的过程（Kawai et al.，1988；Tuor et al.，1992），这也导致了醌基团相对于土壤腐殖质的酚基团的相对富集。除了酚类结构的氧化转化的贡献之外，考虑到醌官能团广泛分布在微生物细胞膜中，土壤腐殖质中的醌基团也可以衍生自微生物坏死部分（Newman and Kolter，2000；Ward et al.，2004）。即使在这种情况下，土壤腐殖质的氧化降解也可能导致苯酚基团优于醌基团降解（Aeschbacher et al.，2012），这可能是由于苯酚基团在氧化环境中相对于醌基团更易于氧化（Rimmer and Smith，2009；Rimmer，2006，2011）。这也导致了在土壤氧化过程中醌基团相对

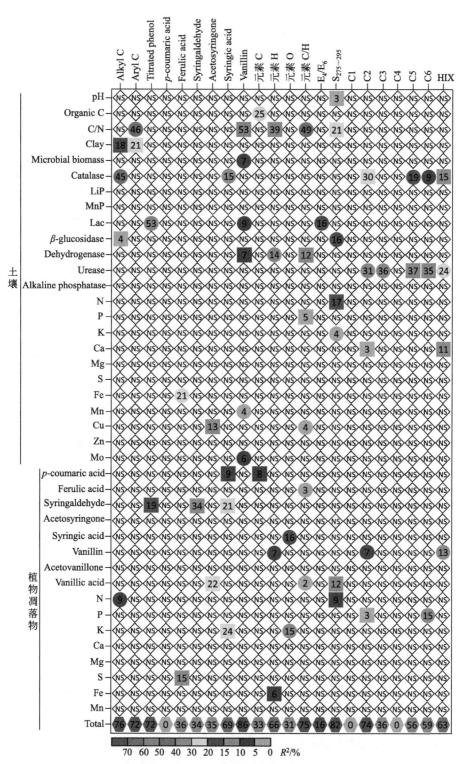

图 4-8 根据逐步多元线性回归得到土壤和植物凋落物指数来反映纬度梯度土壤胡敏酸物理化学性质的指标（见书后彩图 10）

第 4 章 土壤腐殖质电子转移能力对增温的响应

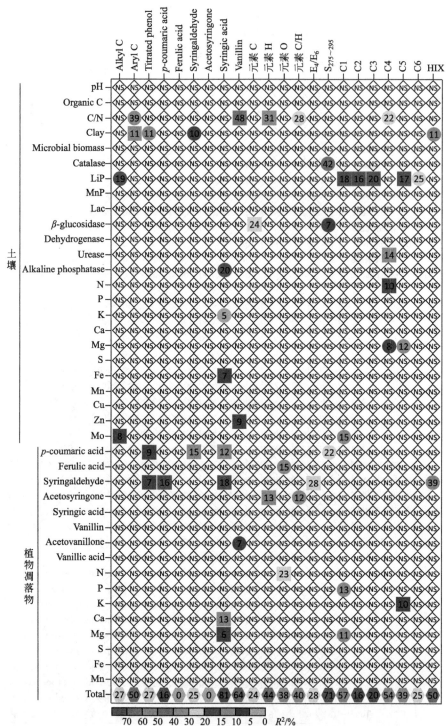

图 4-9 根据逐步多元线性回归得到土壤和植物凋落物指数来反映纬度梯度土壤富里酸物理化学性质的指标（见书后彩图 11）

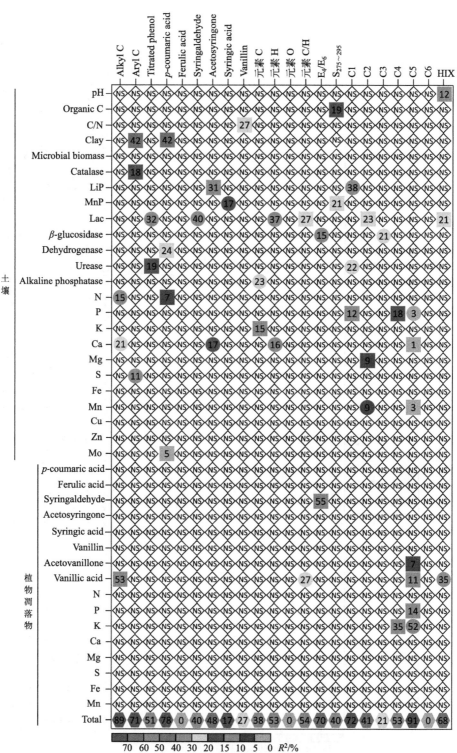

图 4-10 根据逐步多元线性回归得到土壤和植物凋落物指数来反映海拔梯度
土壤胡敏酸物理化学性质的指标（见书后彩图 12）

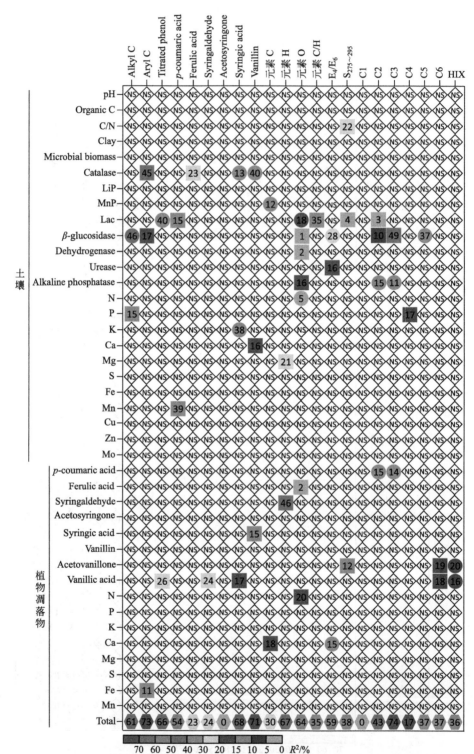

图 4-11 根据逐步多元线性回归得到土壤和植物凋落物指数来反映海拔梯度土壤富力酸物理化学性质的指标（见书后彩图 13）

含量增加和苯酚基团减少的现象。以前的研究已经证明，腐殖质的氧化，包括酚基团氧化转化为醌基团以及苯酚基团的氧化降解是由与木质素氧化相关的酶介导的（Kawai et al.，1988；Tuor et al.，1992）。虽然醌基可以通过羟基自由基与酚基的非生物反应形成（Raghavan and Steenken，1980），但是木质素降解形成的羟基也参与酶反应（Barr et al.，1992；Forney et al.，1982；Gomez-Toribio et al.，2009；Grinhut et al.，2011）。因此，上述酶活性对电子转移能力相关的物理化学性质的影响可以通过调节土壤中腐殖质的氧化转化或降解来表征。

土壤 C/N 值通常取决于土壤有机质的氧化，因为有机物降解可以以二氧化碳的形式释放碳（Swift et al.，1979）。我们的研究结果表明，土壤 C/N 值是芳基 C 和 C/H 值变化的正向预测因子，而与纬向梯度下土壤胡敏酸和富里酸中香草醛和 H 的变化呈负相关（图4-8和图4-9），从而证实了上述观点，即腐殖质的电子转移能力相关的物理化学性质的变化与土壤中腐殖质的氧化直接相关。

跨越纬度梯度的土壤中的胡敏酸和富里酸的芳基 C 和跨越海拔梯度的土壤中的胡敏酸呈显著正相关（图4-8和图4-10），这可能是由于优先吸收芳基 C 的腐殖质在小粒径团聚体中，相比于大粒径团聚体有机物质可以更好地保护免受微生物侵袭。

黄土-古土壤序列的土壤中腐殖质含量随着沉积深度或沉积时间的延长而逐渐降低（表4-4），而与腐殖质的电子转移能力相关的大部分物理化学性质与黄土-古土壤序列的沉积深度无关（图4-12）。这表明沉积时间并没有显著影响腐殖质的化学结构，这可能与黄土-古生物系列序列中腐殖质的非优先降解有关。确定真实的土壤 C/N 比值和黄土古生物序列的微生物活动以及它们如何反映相应地质时间的土壤环境条件是困难的，因为这些参数可能在沉积期间发生变化。因此，需要进一步的研究来探讨土壤 C/N 值和与木质素氧化有关的微生物活性显著影响纬度和海拔梯度上土壤腐殖质的固有物理化学性质的结果是否也适用于黄土-古土壤序列。

表4-4 不同海拔梯度、维度梯度和黄土-古土壤序列土壤样品的胡敏酸和富里酸的含量

单位：g/kg

维度梯度			海拔梯度			黄土-古土壤序列		
采样地点	HA	FA	采样地点	HA	FA	采样地点	HA	FA
1	0.86	1.42	1	1.46	0.94	S0-1	0.89	0.95
2	1.37	1.39	2	1.48	1.10	S0-2	0.78	0.64

续表

维度梯度			海拔梯度			黄土-古土壤序列		
采样地点	HA	FA	采样地点	HA	FA	采样地点	HA	FA
3	1.10	1.24	3	1.60	1.37	L1-1	0.39	0.41
4	1.03	1.19	4	1.16	0.74	L1-2	0.34	0.44
5	1.64	1.82	5	1.48	1.05	L1-3	0.33	0.41
6	2.33	1.77	6	1.49	0.95	L1-4	0.34	0.48
7	0.75	0.54	7	0.92	1.00	S1-1	0.64	0.52
8	1.64	1.36	8	0.94	0.71	S1-2	0.87	0.85
9	0.94	1.32	9	1.25	1.12	S1-3	0.77	0.60
10	2.16	1.60	10	1.62	1.00	L2-1	0.31	0.53
11	2.77	1.42	11	1.11	0.94	L2-2	0.15	0.30
12	2.73	1.47	12	1.21	1.36	L2-3	0.17	0.22
13	1.52	0.83	13	0.89	0.86	L2-4	0.16	0.23
14	1.99	1.08	14	1.42	1.03	S2-1	0.42	0.51
15	2.54	2.05	15	1.57	1.87	S2-2	0.33	0.45
16	2.67	2.06	16	0.84	0.59	S2-3	0.47	0.42
17	1.73	1.20	17	0.74	0.83	L3-1	0.22	0.25
18	2.42	1.86	18	1.24	1.11	L3-2	0.24	0.38
19	1.90	1.65				S3-1	0.29	0.27
20	2.10	1.91				S3-2	0.13	0.09
21	2.07	1.05				S3-3	0.18	0.17
22	0.69	0.86				L4-1	0.18	0.20
23	1.72	1.16				L4-2	0.21	0.27
24	0.82	0.69				L4-3	0.21	0.25
25	1.59	1.49				S4-1	0.16	0.13
26	2.51	1.32				S4-2	0.18	0.18
27	0.75	0.95				S4-3	0.17	0.14

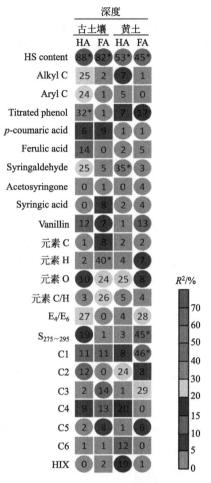

图 4-12 黄土-古生物土壤序列的沉积深度与
土壤胡敏酸物理化学性能的相关性系数（见书后彩图 14）

4.4 土壤和植物凋落物与温度的关系

纬度和海拔梯度的土壤中过氧化氢酶、LiP、MnP 和 Lac 的活性随着温度的升高而显著增加（图 4-13），最有可能的原因是变暖可以通过增加植物初级生产力（Pendall et al., 2004）来增加土壤碳输入量，从而提高微生物活性。此外，跨纬度和海拔梯度的土壤中 C/N 值与温度呈显著负相关（图 4-13），表明天然有机质在温暖条件下氧化能力增强。这与之前的结果一致，即相比于寒冷的生态系统，在温暖的生态系统中，土壤的 C/N 值较低（Callesen et al., 2007）。凋落物木质素衍生的丁香醛与跨纬度和海拔梯度土壤组中的温度呈显著负相关（图 4-13），这可能是因为丁基酚类（如丁香醛）相对于其他木质素衍生的酚类具有较低的热稳定性（Kuo et al.，2008）。此外，土壤 N、P、S 和植物凋落物 K 也与跨纬度或海拔梯度土壤中的温度显著相关（图 4-13）。然而，这些指标并不影响所有土壤组和腐殖质组分中腐殖质的物理化学性质。

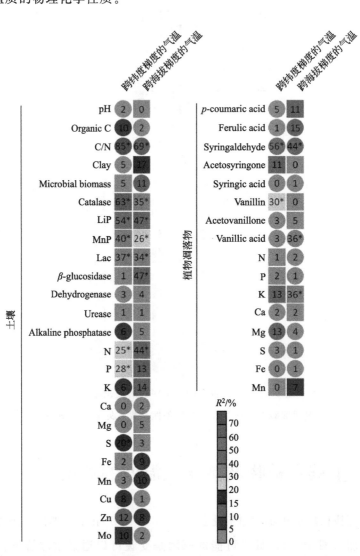

图 4-13 土壤和植物凋落物指数与纬度和海拔梯度的年平均气温的相关系数（见书后彩图 15）

考虑到上述信息，研究结果表明，通过提高与木质素氧化相关的土壤酶活性、升高温度可以提升土壤中腐殖质的氧化转化和降解。当微生物在土壤中与天然有机质和腐殖质易于接触时，随着温度的升高，氧化转化和降解速率的增加与 Arrhenius 速率定律一致（Davidson and Janssens，2006；Friedlingstein et al.，2006）。另外，当天然有机质和腐殖质在物理化学上免受土壤微生物的侵袭时，虽然 Arrhenius 动力学此时并不适用，但升高的温度可能会部分破坏保护机理并暴露底物，从而在土壤中间接促进氧化转化和降解天然有机质和腐殖质（Conant et al.，2011；von Lützow and Kögel-Knabner，2009）。通过升高温度促进腐殖质的氧化导致出现电子供体基团（如滴定酚和特定木质素衍生酚）被逐渐氧化成电子受体醌基团或优先降解的现象，从而增强了土壤中腐殖质的电子接受能力，并降低了腐殖质的电子供给能力。

总体而言，本书首次全面评估了基于通过不同的地质周期的冰期-间冰期循环地带、沿着自然温度梯度的纬度和海拔横断面土壤中的有机物的氧化还原性质（本书中用碱提取腐殖质）对自然温度变化的响应。我们的研究结果表明，随着温度的升高，土壤腐殖质的电子接受能力增加，而电子供给能力减小。结果表明，电子接受能力与电子供给能力的比例可以表征土壤中天然有机质和腐殖质氧化转化和降解的程度。鉴于天然有机质和腐殖质的氧化还原性质对生物地球化学过程的重要性（Klüpfel et al.，2014；Piepenbrock et al.，2014；Piepenbrock and Kappler，2013），这项工作有助于更好地了解在温度升高的背景下涉及土壤有机质的氧化还原性质的生物地球化学过程。

参考文献

Aeschbacher M, Graf C, Schwarzenbach R P, et al, 2012. Antioxidant properties of humic substances. Environmental Science & Technology, 46 (9)：4916-4925.

Aeschbacher M, Sander M, Schwarzenbach R P, 2010. Novel electrochemical approach to assess the redox properties of humic substances. Environmental Science & Technology, 44 (1)：87-93.

Alberts J J, Schindler J E, Miller R W, et al, 1974. Elemental mercury evolution mediated by humic acid. Science, 184 (4139)：895-897.

An Z, Liu T, Lu Y, et al, 1990. The long-term paleomonsoon variation recorded by the loess-paleosol sequence in Central China. Quaternary International, 7-8：91-95.

Barr D P, Shah M M, Grover T A, et al, 1992. Production of hydroxyl radical by lignin peroxidase from *Phanerochaete chrysosporium*. Archives of Biochemistry and Biophysics, 298 (2)：480-485.

Benz M, Schink B, Brune A, 1998. Humic acid reduction by *Propionibacterium freudenreichii* and other fermenting bacteria. Applied and Environmental Microbiology, 64 (11)：

4507-4512.

Borch T, Kretzschmar R, Kappler A, et al, 2010. Biogeochemical redox processes and their impact on contaminant dynamics. Environmental Science and Technology, 44 (1): 15-23.

Callesen I, Rauland-Rasmussen K, Westman C J, et al, 2007. Nitrogen pools and C: N ratios in well-drained Nordic Forest soils related to climate and soil texture. Boreal Environment Research, 12 (6): 681-692.

Cervantes F J, Bok F A M D, Duong-Dac T, et al, 2002. Reduction of humic substances by halorespiring, sulphate-reducing and methanogenic microorganisms. Environmental Microbiology, 4 (1): 51-57.

Chen J, Gu B, Royer R A, et al, 2003. The roles of natural organic matter in chemical and microbial reduction of ferric iron. Science of the Total Environment, 307 (1-3): 167-178.

Conant R T, Ryan M G, Gren G I, et al, 2011. Temperature and soil organic matter decomposition rates-synthesis of current knowledge and a way forward. Global Change Biology, 17 (11): 3392-3404.

Davidson E A, Janssens I A, 2006. Temperature sensitivity of soil carbon decomposition and feedbacks to climate change. Nature, 440 (7081): 165-173.

Dunnivant F M, Schwarzenbach R P, Macalady D L, 1992. Reduction of substituted nitrobenzenes in aqueous solutions containing natural organic matter. Environmental Science & Technology, 26 (11): 2133-2141.

Einsiedl F, Mayer B, Schäfer T, 2008. Evidence for incorporation of H_2S in groundwater fulvic acids from atable isotope ratios and sulfur K-edge X-ray absorption near edge structure spectroscopy. Environmental Science & Technology, 42 (7): 2439-2444.

Forney L J, Reddy C A, Tien M, et al, 1982. The involvement of hydroxyl radical derived from hydrogen peroxide in lignin degradation by the white rot fungus Phanerochaete chrysosporium. Journal of Biological Chemistry, 257 (19): 11455-11462.

Friedlingstein P, Cox P, Betts R, et al, 2006. Climate-carbon cycle feedback analysis: Results from the (CMIP)-M-4 model intercomparison. Journal of Climate, 19 (14), 3337-3353

Gomez-Toribio V, Garcia-Martin A B, Martinez M J, et al, 2009. Induction of extracellular hydroxyl radical production by white-rot fungi through quinone redox cycling. Applied and Environmental Microbiology, 75 (12): 3944-3953.

Grinhut T, Salame T M, Chen Y, et al, 2011. Involvement of ligninolytic enzymes and Fenton-like reaction in humic acid degradation by Trametessp. Applied Microbiology and Biotechnology, 91 (4): 1131-1140.

Gu B, Chen J, 2003. Enhanced microbial reduction of Cr(Ⅵ) and U(Ⅵ) by different natural organic matter fractions. Geochimica et Cosmochimica Acta, 67 (19): 3575-3582.

Han W X, Fang J Y, Reich P B, et al, 2011. Biogeography and variability of eleven mineral elements in plant leaves across gradients of climate, soil and plant functional type in China. Ecology

Letters, 14 (8): 788-796.

Helburn R S, Maccarthy P, 1994. Determination of some redox properties of humic acid by alkaline ferricyanide titration. Analytica Chimica Acta, 295 (3): 263-272.

Hernández-Montoya V, Alvarez L H, Montes-Morán M, et al, 2012. Reduction of quinone and non-quinone redox functional groups in different humic acid samples by Geobacter sulfurreducens. Geoderma, 183-184: 25-31.

Herzsprung P, von Tümpling W, Hertkorn N, et al, 2012. Variations of DOM quality in inflows of a drinking water reservoir: linking of van krevelen diagrams with EEMF spectra by rank correlation. Environmental Science & Technology, 46 (10): 5511-5518.

Imsande J, 1999. Iron-sulfur clusters: Formation, perturbation, and physiological functions. Plant Physiology and Biochemistry, 37 (2): 87-97.

Kappler A, Haderlein S B, 2003. Natural organic matter as reductant for chlorinated aliphatic pollutants. Environmental Science & Technology, 37 (12): 2714-2719.

Kawai S, Umezawa T, Higuchi T, 1988. Degradation mechanisms of phenolic β-1 lignin substructure model compounds by laccase of coriolus versicolor. Archives of Biochemistry and Biophysics, 262 (1): 99-110.

Kiem R, Koögel-Knabner I, 2003. Contribution of lignin and polysaccharides to the refractory carbon pool in C-depleted arable soils. Soil Biology and Biochemistry, 35 (1): 101-118.

Kindler R, Miltner A, Richnow H H, et al, 2006. Fate of gram-negative bacterial biomass in soil-mineralization and contribution to SOM. Soil Biology and Biochemistry, 38 (9): 2860-2870.

Kirk T K, Farrell R L, 1987. Enzymatic "Combustion": The microbial degradation of lignin. Annual Review of Microbiology, 41 (1): 465-501.

Kleber M, Johnson M G, 2010. Advances in understanding the molecular structure of soil organic matter: Implications for interactions in the environment. Advances in Agronomy, 106: 77-142.

Klüpfel L, Piepenbrock A, Kappler A, et al, 2014. Humic substances as fully regenerable electron acceptors in recurrently anoxic environments. Nature Geoscience, 7 (3): 195-200.

Kukla G, 1990. Magnetic susceptibility record of Chinese Loess. Transactions of the Royal Society of Edinburgh Earth Sciences, 81 (4): 263-288.

Kuo L J, Louchouarn P, Herbert B E, 2008. Fate of CuO derived lignin oxidation products during plant combustion: Application to the evaluation of char input to soil organic matter. Organic Geochemistry, 39 (11): 1522-1536.

Lehmann J, Kleber M, 2015. The contentious nature of soil organic matter. Nature, 528 (7580): 60-68.

Leonowicz A, Cho N S, Luterek J, et al, 2001. Fungal laccase: Properties and activity on lignin. Journal of Basic Microbiology, 41 (3-4): 185-227.

Liang C, Balser T C, 2008. Preferential sequestration of microbial carbon in subsoils of a glacial-landscape toposequence, Dane County, WI, USA. Geoderma, 148 (1): 113-119.

Lovley D R, Coates J D, Blunt-Harris E L, et al, 1996. Humic substances as electron acceptors for microbial respiration. Nature, 382 (6590): 445-448.

Lu H Y, Wu N Q, Liu K B, et al, 2007. Phytoliths as quantitative indicators for the reconstruction of past environmental conditions in China II: Palaeoenvironmental reconstruction in the Loess Plateau. Quaternary Science Reviews, 26 (5-6): 759-772.

Marsili E, Baron D B, Shikhare I D, et al, 2008. Shewanella secretes flavins that mediate extracellular electron transfer. Proceedings of the National Academy of Sciences, 105 (10): 3968-3973.

Nakayasu K, Fukushima M, Sasaki K, 1999. Comparative studies of the reduction behavior of chromium-(VI) by humic substances and their precursors. Environmental Toxicology and Chemistry, 18 (6): 1085-1090.

Newman D K, Kolter R, 2000. A role for excreted quinones in extracellular electron transfer. Nature, 405 (6782): 94-97.

Ohno T, 2002. Fluorescence inner-filtering correction for determining the humification index of dissolved organic matter. Environmental Science & Technology, 36 (4): 742-746.

Okamoto A, Hashimoto K, Nealson K H, et al, 2013. Rate enhancement of bacterial extracellular electron transport involves bound flavin semiquinones. Proceedings of the National Academy of Sciences, 110 (19): 7856-7861.

Palmer T, Sargent F, Berks B C, 2005. Export of complex cofactor-containing proteins by the bacterial Tat pathway. Trends in Microbiology, 13 (4): 175-180.

Parmesan C, Yohe G, 2003. A globally coherent fingerprint of climate change impacts across natural ecosystems. Nature, 421 (6918): 37-42.

Pendall E, Bridgham S, Hanson P J, et al, 2004. Below-ground process responses to elevated CO_2 and temperature: A discussion of observations, measurement methods, and models. New Phytologist, 162 (2): 311-322.

Peterse F, Prins M A, Beets C J, et al, 2011. Decoupled warming and monsoon precipitation in East Asia over the last deglaciation. Earth and Planetary Science Letters, 301 (1-2): 0-264.

Piepenbrock A, Kappler A, 2013. Humic substances and extracellular electron transfer. Microbial Metal Respiration, 107-128.

Piepenbrock A, Schroeder C, Kappler A, 2014. Electron transfer from humic substances to biogenic and abiogenic Fe(III) oxyhydroxide minerals. Environmental Science & Technology, 48 (3): 1656-1664.

Raghavan N V, Steenken S, 1980. Electrophilic reaction of the hydroxyl radical with phenol. Determination of the distribution of isomeric dihydroxycyclohexadienyl radicals. Journal of the American Chemical Society, 102 (10): 3495-3499.

Ratasuk N, Nanny M A, 2007. Characterization and quantification of reversible redox sites in humic substances. Environmental Science & Technology, 41 (22), 7844-7850.

Rimmer D L, 2006. Free radicals, antioxidants, and soil organic matter recalcitrance. European Journal of Soil Science, 57 (2): 91-94.

Rimmer D L, Abbott G D, 2011. Phenolic compounds in NaOH extracts of UK soils and their contribution to antioxidant capacity. European Journal of Soil Science, 62 (2): 285-294.

Rimmer D L, Smith A M, 2009. Antioxidants in soil organic matter and in associated plant materials. European Journal of Soil Science, 60 (2): 170-175.

Roden E E, Kappler A, Bauer I, et al, 2010. Extracellular electron transfer through microbial reduction of solid-phase humic substances. Nature Geoscience, 3 (6): 417-421.

Scott D T, Mcknight D M, Blunt-Harris E L, et al, 1998. Quinone moieties act as electron acceptors in the reduction of humic substances by humics-reducing microorganisms. Environmental Science & Technology, 32 (19): 2984-2989.

Simpson A J, Simpson M J, Smith E, et al, 2007. Microbially derived inputs to soil organic matter: are current estimates too low? Environmental Science & Technology, 41 (23): 8070-8076.

Stevenson F J, 1994. Humus Chemistry. New York: John Wiley & Sons.

Struyk Z, Sposito G, 2001. Redox properties of standard humic acids. Geoderma, 102 (3-4): 329-346.

Swift M J, Heal O W, Anderson J M, 1979. Decomposition in terrestrial ecosystems. Studies in Ecology, 5 (14): 2772-2774.

Swift R S, 1996. Organic matter characterization. In Methods of Soil Analysis, Part 3. Chemical Methods, Soil Science Society of America, 1018-1021.

Thevenot M, Dignac M F, Rumpel C, 2010. Fate of lignins in soils: A review. Soil Biology and Biochemistry, 42 (8): 1200-1211.

Tien M, Kirk T K, 1983. Lignin-degrading enzyme from the hymenomycete *Phanerochaete chrysosporium* Burds. Science, 221 (4611): 661-663.

Tuor U, Wariishi H, Schoemaker H E, et al, 1992. Oxidation of phenolic arylglycerol β-aryl ether lignin model compounds by manganese peroxidase from Phanerochaete chrysosporium: Oxidative cleavage of an α-carbonyl model compound. Biochemistry, 31 (21): 4986-4995.

Turick C E, Tisa L S, Caccavo F, 2002. Melanin production and use as a soluble electron shuttle for Fe(Ⅲ) oxide reduction and as a terminal electron acceptor by Shewanella algae BrY. Applied and Environmental Microbiology, 68 (5): 2436-2444.

Van der Zee F R, Cervantes F J, 2009. Impact and application of electron shuttles on the redox (bio) transformation of contaminants: A review. Biotechnology Advances, 27 (3): 256-277.

Von Luützow M, Koögel-Knabner I, 2009. Temperature sensitivity of soil organic matter decomposition-what do we know? Biology and Fertility of Soils, 46 (1): 1-15.

Walpen N, Schroth M H, Sander M, 2016. Quantification of phenolic antioxidant moieties in dissolved organic matter by flow-injection analysis with electrochemical detection. Environmental Science & Technology, 50 (12): 6423-6432.

Ward M J, Fu Q S, Rhoads K R, et al, 2004. A derivative of the menaquinone precursor 1,4-dihydroxy-2-naphthoate is involved in the reductive transformation of carbon tetrachloride by aerobically grown *Shewanella oneidensis* MR-1. Applied Microbiology and Biotechnology, 63 (5): 571-577.

White G F, Shi Z, Shi L, et al, 2013. Rapid electron exchange between surface-exposed bacterial cytochromes and Fe(Ⅲ) minerals. Proceedings of the National Academy of Sciences, 110 (16): 6346-6351.

Wittbrodt P R, Palmer C D, 1997. Reduction of Cr(Ⅵ) by soil humic acids. European Journal of Soil Science, 48 (1): 151-162.

第5章 土壤腐殖质竞争性抑制甲烷生成对增温的响应

以一百年为时间尺度的预测中，甲烷（CH_4）引起全球变暖的潜能值是二氧化碳（CO_2）的25倍以上（IPCC，2007），因此未来大气CH_4浓度的微小改变便会对气候变化产生重大影响。CH_4对工业化时期以来人为导致的辐射强度贡献了约20%，使其成为继CO_2后的第二大温室气体（IPCC，2007）。大气CH_4的变化是人类和自然资源共同变化导致的结果（Bousquet et al.，2006）。处于厌氧状态的水稻田和湿地，分别是CH_4排放最大的人为源和天然源（IPCC，2007；Bridgham et al.，2013），二者共同贡献了全球CH_4排放总量的1/3（Bridgham et al.，2013）。因此，了解水稻田和自然湿地中CH_4的生物地球化学动力学过程具有重要意义。

在诸如水稻田和湿地等厌氧环境中，CH_4主要是通过氢营养型和乙酸营养型两种途径生成（Whiticar et al.，1986；Whiticar，1999）。在氢营养型途径中，H_2是产甲烷菌唯一可以利用的终端发酵产物，其可以通过利用CO_2作为电子受体被氧化成CH_4（Bridgham et al.，2013）。在乙酸营养型途径中，乙酸通过裂解反应形成CH_4和CO_2（Bridgham et al.，2013）。越来越多的证据表明，这两种途径中的H_2和乙酸在甲烷生成的最后一步的代谢中，也可以被许多微生物在厌氧呼吸时利用腐殖质作为末端电子受体（TEA）所利用（Cervantes et al.，2000；Keller and Bridgham，2007；Blodau and Deppe，2012；Miller et al.，2015）。因此，根据反应热力学原理，更容易发生的腐殖质微生物还原反应可能会竞争性地抑制CH_4的生成（Klüpfel et al.，2014）。而且，由于水稻田和湿地这两个生态系统中的土壤有机质主要是由腐殖质构成，所以其腐殖质微生物还原反应对CH_4生成的竞争性抑制可能会显得尤为重要（Stevenson，1994）。

甲烷生成过程和腐殖质微生物还原反应都受到微生物活性的影响（Singh et

al.，2010；Martinez et al.，2013），在很大程度上取决于环境条件。作为全球气候变化的主要特征之一，气候变暖是一个无可争议和不可避免的事实（IPCC，2013），并被认为是影响生态系统微生物活性的重要因素（Singh et al.，2010）。虽然甲烷生成过程和腐殖质微生物还原反应都与气候变暖存在密切的联系，但是关于腐殖质微生物还原反应对 CH_4 生成的竞争性抑制是否与气候变暖存在内在关系目前尚不清楚。显然，这种关系是值得我们去研究的，因为它可以帮助我们理解厌氧环境中 CH_4 动态过程和制定相关管理策略以应对气候变暖背景下厌氧生态系统中 CH_4 排放。

在本书中，不同温度条件下对添加不同腐殖质的水稻田土壤和湿地土壤进行厌氧培养。同时，在相应的温度条件下培养添加水的水稻田土壤和湿地土壤作为对照。在整个土壤厌氧培养过程中，监测 CH_4 的生成。本书的主要目标有两个：一是验证添加腐殖质对厌氧条件下 CH_4 生成的影响；二是检验温度是否会影响腐殖质微生物还原反应竞争性地抑制 CH_4 生成。

5.1　添加腐殖质对厌氧条件下甲烷和二氧化碳生成的影响

水稻田土壤和湿地土壤在温育条件下，产 CH_4 和 CO_2 量呈线性增长（图 5-1）。随着温育时间的增加，添加腐殖质中的 CH_4 浓度增长速度比对照组慢；而与对照组相比，添加腐殖质组中的 CO_2 排放量却没有发生改变（图 5-1）。通过比较添加腐殖质组和对照组之间 CH_4 和 CO_2 的产生速率，进一步验证了此结果（图 5-2）。我们的研究结果表明，腐殖质有潜力在厌氧环境中抑制甲烷的生成（Blodau and Deppe，2012；Miller et al.，2015）。

腐殖质从微生物接受电子的过程可以竞争性地抑制其他末端电子受体（TEAs）的还原，包括产甲烷条件下的 CO_2（Keller et al.，2009；Blodau and Deppe，2012；Bridgham et al.，2013）。在本次研究中，首先是土壤腐殖酸（ES-HA）显示出最高的电子接受能力，其次是泥炭腐殖酸（PPHA）和河腐殖酸（SRHA）(Klüpfel et al.，2014）。电子接受容量的顺序与厌氧系统中抑制产 CH_4 的腐殖质的强度相同（图 5-1 和图 5-3）。鉴于腐殖质的微生物还原反应与腐殖质的电子接受能力呈正相关（Lovley et al.，1996），可以推断，相比与对照组，添加腐殖质组中产 CH_4 率较低的原因可能与腐殖质的微生物还原反应有关。

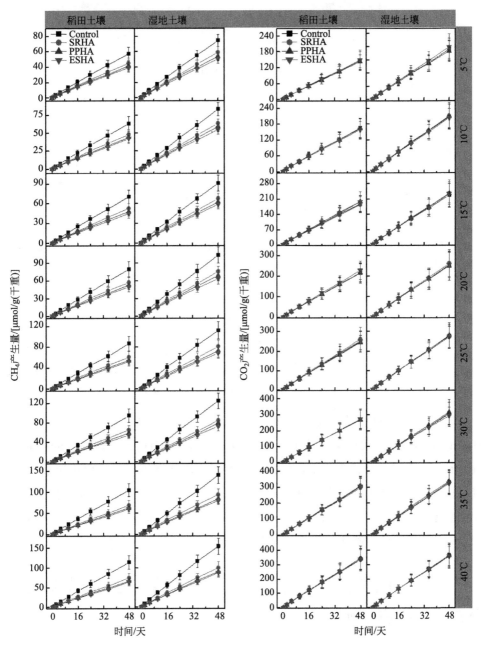

图 5-1 水稻田与湿地土壤在不同温度、添加不同腐殖质及缺氧培养下对 CH_4 和 CO_2 产生过程的影响

Control—对照组；ESHA—土壤腐殖酸；PPHA—泥炭腐殖酸；SRHA—河腐殖酸

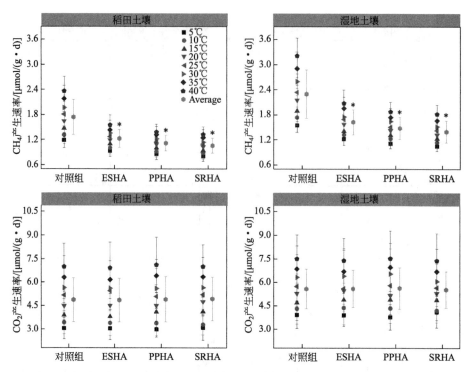

图 5-2 水稻田与湿地土壤在不同温度缺氧培养下对 CH_4 和 CO_2 产生速率的影响

注：星号（*）表示添加腐殖质处理与对照之间存在显著差异（$p<0.05$）。

此外，考虑添加天然有机物质可能会引起 CH_4 厌氧氧化，因此，腐殖质引起的 CH_4 厌氧氧化也可能会导致添加腐殖质组中较低的 CH_4 产生速率（Smemo and Yavitt，2007，2011）。理论上，相比于对照组，添加腐殖质组中由腐殖质引起的 CH_4 厌氧氧化会导致更多的 CO_2 生成，或 CO_2 释放速率的增大，但在本书中却没有发生这种现象（图 5-1 和图 5-2）。这也间接地表明厌氧条件下 CH_4 产生速率的降低是由腐殖质微生物还原而不是由腐殖质诱导的 CH_4 氧化导致的。

5.2 腐殖质微生物还原反应竞争性抑制甲烷生成对温度升高的响应

腐殖质的竞争性抑制产 CH_4 的强度与培育温度呈显著正相关（图 5-3），表明温度是提高腐殖质微生物还原反应的竞争性抑制产甲烷的重要因素。需要注意的是，腐殖质的竞争性抑制产 CH_4 强度和培育温度之间的良好相关性适用于水稻田土壤和湿地土壤（图 5-3），表明温度对腐殖质微生物还原反应的竞争性抑制产甲

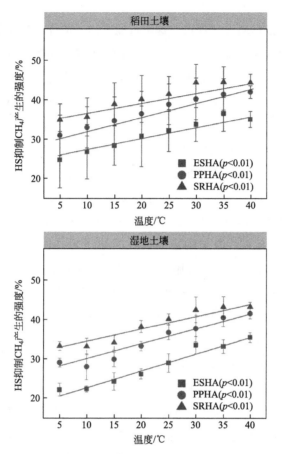

图 5-3 腐殖质抑制水稻田和湿地土壤甲烷生成随温度的变化

烷的影响在不同生态系统中是普遍的。

虽然在不同培育温度下，ESHA、PPHA 和 SRHA 具有不同的竞争性抑制产 CH_4 强度，但是随着培育温度的升高，它们的竞争性抑制产 CH_4 强度变化产生了类似的线性斜率（图 5-3）。这些结果可能是由于 ESHA、PPHA 和 SRHA 在厌氧培育系统中使用相同的氧化还原活性官能团来接受电子。喹诺酮部分是腐殖质微生物还原反应中最重要的电子接受官能团（Scott et al.，1998；Aeschbacher et al.，2011），尽管非醌类芳族结构和络合金属离子也被认为是可能的氧化还原活性位点（Struyk and Sposito，2001；Chen et al.，2003；Einsiedl et al.，2008）。因此，我们可以推断，ESHA、PPHA 和 SRHA 中的醌基部分在厌氧培育系统中的竞争性抑制产 CH_4 过程中发挥了最大的作用。

鉴于产甲烷和腐殖质微生物还原反应都是类似于酶介导的生物化学反应

(Thauer，1998；Martinez et al.，2013)，这两个过程的反应速率预计会随着短期温度升高而增加，根据动力学理论，这很可能是由于酶介导反应的速率会随着温度的升高而明显增加（Arrhenius，1889）。产甲烷和腐殖质微生物还原反应由不同的酶或蛋白质系统参与，因此需要不同的活化能才可以进行。考虑到生物化学反应的温度敏感性对活化能的动力学依赖性，可能导致产甲烷和腐殖质微生物还原反应具有不同的温度敏感性（Davidson and Janssens，2006）。笔者认为参与产甲烷过程的酶活性可能比涉及腐殖质微生物还原反应的酶活性具有更低的温度敏感性，这部分地导致了温度升高使腐殖质微生物还原反应的竞争性抑制产 CH_4 强度的升高。

厌氧条件下产 CH_4 的相关过程，包括氢营养型甲烷化、乙酸发酵型甲烷化、产乙酸作用、二次发酵和同型产乙酸作用都被认为是胞内代谢过程（Thauer，1998）。相比之下，腐殖质微生物还原反应则被认为是胞外呼吸过程（Lovley et al.，1996），并且这个过程由与细胞表面上的电子传递蛋白结合的辅酶因子来介导（Clarke et al.，2011）。细胞膜可以通过隔离细胞质与胞外环境创造出一个相对独立的胞内环境。因此，与腐殖质微生物还原反应这类胞外反应相比，微生物细胞内的产 CH_4 过程可能不太容易升高温度。这可能是温度升高时腐殖质微生物还原反应的竞争性产 CH_4 抑制得到增强的另一个原因。

传统意义上腐殖质被认为是植物和微生物残留物被微生物降解形成的高分子有机化合物，并且是构成陆地和水生系统中天然有机质的主要部分（Stevenson，1994）。与将腐殖质定义为结构复杂的大分子物质相反，一个新的模型表明，腐殖质并不是由单一的大分子组成的，而是由许多分子量较小和结构更简单的分子靠化学键组合在一起的化合物（Sutton and Sposito，2005）。然而，重要的是要认识到氧化还原活性官能团是腐殖质和天然有机质的一部分。因此，在真实的生态系统中存在温度对由腐殖质或天然有机物产生的产甲烷抑制的积极影响便是可以合理预计的。基于真实纬度和高度横断面的研究结果表明，土壤中腐殖质接受电子官能团数量会随着温度的升高而增加（Tan et al.，2017）。因此，在真实变暖的环境中，除了通过本书中发现的途径之外，温度升高还可以通过增加腐殖质中接受电子基团的数量来增强通过腐殖质微生物还原反应产生的产 CH_4 竞争性抑制。

5.3　环境意义

虽然关于腐殖质的产甲烷抑制与温度之间联系的实际机制需要进一步探究，但关于全球变暖对厌氧条件下腐殖质微生物还原反应对产甲烷的竞争性抑制作用的研究有以下一些环境方面的应用。第一，研究结果表明，在人为气候变暖的背景下，作为末端电子受体的腐殖质将在缓解湿地和水稻田 CH_4 排放方面发挥越来越重要的作用。第二，几种模型已被用于预测未来 CH_4 排放趋势（Potter，1997；Fumoto et al.，2010；Petrescu et al.，2010；Minamikawa et al.，2014），建议应将腐殖质抑制产甲烷的积极影响纳入模型中，以提高模型的精度。第三，从减少 CH_4 排放的角度来看，研究结果可以帮助战略性地规范由淡水资源稀缺、耕地和劳动力资源共同决定的全球水稻田种植面积和稻米季节性种植模式的变化。

参考文献

Aeschbacher M, Vergari D, Schwarzenbach R P, et al, 2011. Electrochemical analysis of proton and electron transfer equilibria of the reducible moieties in humic acids. Environmental Science & Technology, 458 (19): 8385-8394.

Arrhenius S, 1889. Über die reaktionsgeschwindigkeit bei der inversion von rohrzucker durch säuren. Zeitschrift für physikalische Chemie, 4 (1): 226-248.

Blodau C, Deppe M, 2012. Humic acid addition lowers methane release in peats of the Mer Bleue bog, Canada. Soil Biology and Biochemistry, 52: 96-98.

Bousquet P, Ciais P, Miller J B, et al, 2006. Contribution of anthropogenic and natural sources to atmospheric methane variability. Nature, 443 (7110): 439-443.

Bridgham S D, Cadillo-Quiroz H, Keller J K, et al, 2013. Methane emissions from wetlands: Biogeochemical, microbial, and modeling perspectives from local to global scales. Global Change Biology, 19 (5): 1325-1346.

Cervantes F J, Velde S V D, Lettinga G, et al, 2000. Competition between methanogenesis and quinone respiration for ecologically important substrates in anaerobic consortia. Fems Microbiology Ecology, 34 (2): 161-171.

Chen J, Gu B, Royer R A, et al, 2003. The roles of natural organic matter in chemical and microbial reduction of ferric iron. Science of the Total Environment, 307 (1-3): 167-178.

Clarke T A, Edwards M J, Gates A J, et al, 2011. Structure of a bacterial cell surface decaheme electron conduit. Proceedings of the National Academy of Sciences, 108 (23): 9384-9389.

Davidson E A, Janssens I A, 2006. Temperature sensitivity of soil carbon decomposition and feedbacks to climate change. Nature, 440 (7081): 165-173.

Einsiedl F, Mayer B, Schäfer T, 2008. Evidence for incorporation of H_2S in groundwater fulvic acids from atable isotope ratios and sulfur K-edge X-ray absorption near edge structure spectroscopy. Environmental Science & Technology, 42 (7): 2439-2444.

Fumoto T, Yanagihara T, Saito T, et al, 2010. Assessment of the methane mitigation potentials of alternative water regimes in rice fields using a process-based biogeochemistry model. Global Change Biology, 16 (6): 1847-1859.

IPCC (Intergovernmental Panel on Climate Change), 2007. Changes in atmospheric constituents and in radiative forcing. //Solomon S, Qin D, Manning M, et al. Climate Change 2007: The Physical Science Basis. Contribution of Working Group I to the Fourth Assessment Report of the Intergovernmental Panel on Climate Change, 129-234. Cambridge University Press, Cambridge, United Kingdom and New York, NY, USA.

IPCC (Intergovernmental Panel on Climate Change), 2013. Summary for policymakers. //Stocker T F, Qin D, Plattner G. -K, et al. Climate Change 2013: The Physical Science Basis. Contribution of working Group I to the Fifth Assessment Report of the Intergovernmental Panel on Climate Change, 3-29. Cambridge University Press, Cambridge and New York, NY, USA.

Keller J K, Bridgham S D, 2007. Pathways of anaerobic carbon cycling across an ombrotrophic-minerotrophic peatland gradient. Limnology and Oceanography, 52 (1): 96-107.

Keller J K, Weisenhorn P B, Megonigal J P, 2009. Humic acids as electron acceptors in wetland decomposition. Soil Biology and Biochemistry, 41 (7): 1518-1522.

Klüpfel L, Piepenbrock A, Kappler A, et al, 2014. Humic substances as fully regenerable electron acceptors in recurrently anoxic environments. Nature Geoscience, 7 (3): 195-200.

Lovley D R, Coates J D, Blunt-Harris E L, et al, 1996. Humic substances as electron acceptors for microbial respiration. Nature, 382 (6590): 445-448.

Martinez C M, Alvarez L H, Celis L B, et al, 2013. Humus-reducing microorganisms and their valuable contribution in environmental processes. Applied and Environmental Microbiology, 97 (24): 10293-10308.

Miller K E, Lai C T, Friedman E S, et al, 2015. Methane suppression by iron and humic acids in soils of the Arctic Coastal Plain. Soil Biology and Biochemistry, 83: 176-183.

Minamikawa K, Fumoto T, Itoh M, et al, 2014. Potential of prolonged midseason drainage for reducing methane emission from rice paddies in Japan: a long-term simulation using the DNDC-Rice model. Biology and Fertility of Soils, 50 (6): 879-889.

Petrescu A M R, Beek L P H V, Huissteden J V, et al, 2010. Modeling regional to global CH_4 emissions of boreal and Arctic wetlands. Global Biogeochemical Cycles, 24, GB4009.

Potter C S, 1997. An ecosystem simulation model for methane production and emission from wetlands. Global Biogeochemical Cycles, 11 (4): 495-506.

Scott D T, Mcknight D M, Blunt-Harris E L, et al, 1998. Quinone moieties act as electron acceptors in the reduction of humic substances by humics-reducing microorganisms. Environmental Science & Technology, 32 (19): 2984-2989.

Singh B K, Bardgett R D, Smith P, et al, 2010. Microorganisms and climate change: Terrestrial feedbacks and mitigation options. Nature Reviews Microbiology, 8 (11): 779-790.

Smemo K A, Yavitt J B, 2007. Evidence for anaerobic CH_4 oxidation in freshwater peatlands. Geomicrobiology, 24 (7): 583-597.

Smemo K A, Yavitt J B, 2011. Anaerobic oxidation of methane: An underappreciated aspect of methane cycling in peatland ecosystems? Biogeosciences, 8 (3): 779-793.

Stevenson F J, 1994. Humus chemistry: Genesis, composition, reactions. John Wiley & Sons.

Struyk Z, Sposito G, 2001. Redox properties of standard humic acids. Geoderma, 102 (3-4): 329-346.

Sutton R, Sposito G, 2005. Molecular structure in soil humic substances: The new view. Environmental Science & Technology, 39 (23): 9009-9015.

Tan W, Xi B, Wang G, et al, 2017. Increased electron-accepting and decreased electron-donating capacities of soil humic substances in response to increasing temperature. Environmental Science & Technology, 51 (6): 3176-3186.

Thauer R K, 1998. Biochemistry of methanogenesis: A tribute to Marjory Stephenson: 1998 Marjory Stephenson Prize Lecture. Microbiology, 144 (9): 2377-2406.

Whiticar M J, Faber E, Schoell M, 1986. Biogenic methane formation in marine and freshwater environments: CO_2 reduction vs. acetate fermentation-Isotope evidence. Geochimica et Cosmochimica Acta, 50 (5): 693-709.

Whiticar M J, 1999. Carbon and hydrogen isotope systematics of bacterial formation and oxidation of methane. Chemical Geology, 161 (1-3): 291-314.

第6章 土壤腐殖质电子转移能力对土地利用变化的响应

6.1 土壤腐殖质电子转移能力对农用地类型的响应

腐殖质是陆地和水生系统中天然有机物的主要部分（Aiken et al., 1985; Stevenson, 1994）。在缺氧条件下，腐殖质可以被多种不同的微生物还原，例如硫酸盐还原剂（Cervantes et al., 2002）、铁还原剂（Lovley et al., 1996）和发酵细菌（Benz et al., 1998）。还原的腐殖质可以将电子提供给具有更高氧化还原电位的其他电子受体，如氧化铁和氢氧化物（Lovley et al., 1996; Bauer and Kappler, 2009）以及各种有机和无机污染物，包括氯化物（Kappler and Haderlein, 2003）、硝基苯（Van der Zee and Cervantes, 2009）、U(Ⅵ)（Gu and Chen, 2003）和 Cr(Ⅵ)（Wittbrodt and Palmer, 1997）。因此，腐殖质的电子转移特性可以对环境中氧化还原活性物质的生物地球化学氧化还原过程产生重大影响。

腐殖质的电子转移能力取决于其固有的化学结构。已有研究指出，醌基是腐殖质中的氧化还原活性官能团（Tratnyek and Macalady, 1989; Dunnivant et al., 1992）。Scott 等（1998）研究发现，通过电子自旋共振技术揭示微生物分解腐殖质过程中，醌基充当了最为重要的氧化还原活性官能团。基因证据表明，甲萘醌在分解腐殖质的过程中参与了 *Shewanella* 的电子传输（Newman and Kolter, 2000）。傅里叶变换红外光谱、NMR 光谱和热解-GC-MS 技术的结果也表明，醌基是腐殖质中重要的氧化还原活性官能团（Aeschbacher et al., 2010; Hernández-Montoya et al., 2012）。除喹啉和酚官能团外，含氮和硫的基团［如 1-甲基-2,5-吡咯烷二酮、3-(甲硫基)丙酸、二甲基砜、N-甲基苯胺］和络合金属离子也被认为是腐殖质中氧化还原活性官能团的重要组成部分（Fimmen et al., 2007; Struyk and

Sposito，2001；Einsiedl et al.，2008)。

土壤腐殖质中氧化还原活性官能团的分布和丰度通常取决于腐殖质的微生物分解和转化（Stevenson，1994；Kleber and Johnson，2010）。未来几十年土地管理和土地利用变化将改变陆地生态系统中天然有机物和腐殖质的降解和转化（Shaver et al.，2000），并可能直接影响土壤腐殖质的内在化学结构（Feng et al.，2008；Pisani et al.，2014；Pisani et al.，2015)。因此，有理由预测土壤腐殖质的电子转移能力与土地管理和土地利用变化存在密切的关系。土地管理和土地利用变化对土壤腐殖质电子转移能力的影响机制值得研究，其中涉及土地利用变化的背景下土壤腐殖质的氧化还原功能，对于理解生物地球化学过程具有重要意义。

对中国常州市的不同农用地的土壤进行了采样，使用微生物还原法对土壤中胡敏酸和富里酸的电子转移能力进行量化，以评估土壤胡敏酸和富里酸电子转移能力对农用地类型的响应。这项研究结果为深入了解腐殖质氧化还原特性对不同农用地土壤中有机和无机污染物的迁移、转化和氧化还原转化的作用具有重要意义。

6.1.1 不同农用地类型下的土壤腐殖质电子转移能力

在不同农用地类型下，土壤之间的腐殖质电子转移能力差异显著（$p<0.05$）。在 *S. oneidensis* MR-1 和 *S. putrefaciens* 200 进行的两次接种中，水稻田土壤（PS）中腐殖质电子转移能力最高，其次分别是番茄地土壤（TS）、芹菜白菜地土壤（CCS）、葡萄地土壤（GS）和杨梅地土壤（MRS）(图 6-1)。这些结果表明，农用地类型可以对土壤腐殖质电子转移能力产生重大影响。值得注意的是，在不同的农用地类型中腐殖质电子转移能力与富里酸浓度密切相关（图 6-1），这表明农用地类型对腐殖质电子转移能力的影响与腐殖质的组成无关。

6.1.2 土壤腐殖质化学结构对电子转移能力的影响

采用反映土壤腐殖质化学结构的一系列指标来评估土壤腐殖质理化性质对其电子转移能力的影响。这些指标包括元素组成及其百分比、腐殖化指数（HIX）、荧光光谱、254nm 紫外线吸收率（$SUVA_{254}$）、465nm 和 665nm 的紫外线可见吸收比（E_4/E_6）、光谱面积（$A_{240\sim400}$）和光谱斜率（$S_{250\sim600}$）。通过使用平行因子方

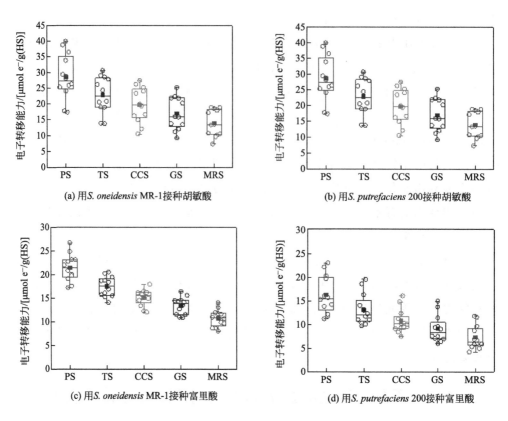

图 6-1 不同农用地类型下土壤腐殖质的电子转移能力

注：不同的小写字母表示显著差异，相同的小写字母表示没有显著差异（$p<0.05$）。

法解析对三维荧光光谱得到的不同组分，即一个类蛋白物质（C1）和三个类腐殖质物质（C2~C4）（图 6-2）。相关性分析表明，土壤腐殖质电子转移能力与其 C/H 值呈显著正相关，与其 E_4/E_6 呈显著负相关，土壤富里酸的电子转移能力与其 C/H 值、HIX 和 C2 呈显著正相关（图 6-3）。研究结果表明，评估腐殖质电子转移能力的化学结构指标与腐殖质结构有关。

通常情况下，天然有机物的高 C/H 和 HIX 表明芳香环中的高聚合（Stevenson, 1994；Kleber and Johnson, 2010；Ohno, 2002）。天然有机物的低 E_4/E_6 很大程度上归因于芳香族碳-碳双键基团的吸收（Kleber and Johnson, 2010）。比较荧光和傅里叶变换离子回旋共振质谱分析的结果表明，类腐殖质荧光通常与芳香结构共变（Herzsprung et al., 2012）。C/H、HIX、C2 和 E_4/E_6 可以用来表示醌型结构。因此，我们的结果表明芳香性在土壤腐殖质中起氧化还原活性基团的作用，支持了现有观点认为土壤腐殖质的醌基是接受

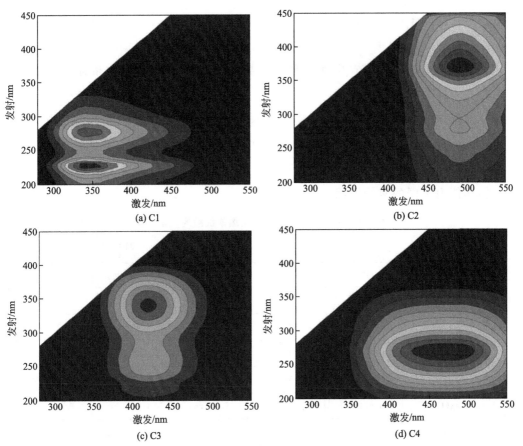

图 6-2 使用平行因子分析确定的荧光成分（见书后彩图 16）

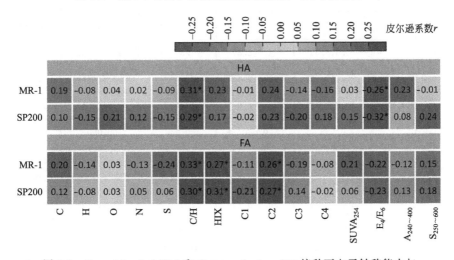

图 6-3 *S. oneidensis* MR-1 和 *S. putrefaciens* 200 接种下电子转移能力与
胡敏酸和富里酸的化学结构因子的相关性（见书后彩图 17）

注：颜色和数字表示相关强度，显著相关（*）在 $p<0.05$ 水平上进行评估。

电子的主要官能团（Lovley et al.，1996；Aeschbacher et al.，2011；Aeschbacher，et al.，2012）。

进一步比较了不同农用地类型下土壤中腐殖质的化学结构指标。结果表明，在不同的农用地类型下，C/H值、HIX和C2与胡敏酸或富里酸的电子转移能力排序相同（表6-1和表6-2）。相反，在不同农用地类型下的土壤中，腐殖质E_4/E_6的排序与腐殖质电子转移能力相反（表6-1）。因此，C/H值、HIX、C2和E_4/E_6的不同是导致不同农用地类型之间土壤腐殖质或富里酸电子转移能力不同的直接原因。

表6-1 不同农用地类型下土壤胡敏酸的化学结构

均值（±SE，$n=3$）

化学结构	PS	TS	CCS	GS	MRS
C/[g/kg(HS)]	468(36)	533(17)	543(25)	467(34)	432(30)
H/[g/kg(HS)]	39(7)	42(5)	52(4)	50(5)	53(6)
O/[g/kg(HS)]	304(14)	309(9)	309(15)	301(17)	301(12)
N/[g/kg(HS)]	31.3(9.4)	23.5(7.2)	30.4(6.8)	28.7(8.1)	30.1(7.5)
S/[g/kg(HS)]	6.4(1.6)	5.8(0.9)	5.7(1.4)	4.8(1.8)	5.0(1.5)
C/H		1.13^a(0.08)	1.02^b(0.06)	0.99^b(0.10)	0.74^b(0.07)
HIX	0.86(0.05)	0.89(0.03)	0.83(0.04)	0.96(0.04)	0.93(0.04)
C1/%	13.0(1.2)	1.15^a(0.06)	14.7(3.4)	7.9(2.5)	7.6(1.4)
C2/%	30.6(3.3)	34.0(2.8)	27.5(5.4)	42.5(6.8)	38.3(5.1)
C3/%	22.4(2.5)	25.4(2.6)	23.1(3.2)	27.5(4.7)	25.3(5.0)
C4/%	34.0(2.2)	31.3(2.9)	34.7(3.1)	22.1(3.4)	28.8(3.3)
$SUVA_{254}$/[L/(m·mg)]	0.055(0.003)	0.060(0.004)	0.046(0.004)	0.050(0.007)	0.051(0.008)
E_4/E_6	3.21^a(0.68)	3.73^b(0.87)	4.31^c(1.01)	4.69^{c,d}(0.96)	5.06^d(0.86)
$A_{240\sim400}$	55.8(2.8)	56.0(3.7)	54.9(3.4)	57.9(3.6)	61.7(2.4)
$S_{250\sim600}$	0.0019(0.0002)	0.0023(0.0001)	0.0017(0.0001)	0.0021(0.0002)	0.0020(0.0001)

注：相同小写字母上标（C/H和HIX）表示没有显著差异，不同小写字母上标表示有显著差异（$p<0.05$）。

表 6-2　不同农用地类型下土壤富里酸的化学结构

均值（±SE，$n=3$）

化学结构	PS	TS	CCS	GS	MRS
C/[g/kg(HS)]	425(25)	449(28)	410(38)	453(39)	479(30)
H/[g/kg(HS)]	55(4)	55(3)	51(8)	52(3)	48(3)
O/[g/kg(HS)]	342(7)	341(9)	320(13)	377(18)	315(5)
N/[g/kg(HS)]	20.1(8.3)	27.0(7.3)	20.9(7.5)	20.1(8.3)	20.5(6.7)
S/[g/kg(HS)]	5.8(0.9)	5.6(1.1)	6.7(0.8)	5.0(1.0)	6.1(0.8)
C/H	0.71[a](0.05)	0.74[b](0.07)	0.75[b](0.08)	0.81[c](0.09)	0.87[d](0.08)
HIX	0.84[a](0.05)	0.88[b](0.02)	0.90[b,c](0.02)	0.93[c,d](0.04)	0.96[d](0.03)
C1/%	24.1(5.3)	22.5(4.2)	23.3(3.4)	25.7(2.9)	21.7(3.0)
C2/%	25.2[a](3.2)	31.2[b](2.2)	33.9[b,c](3.1)	35.1[c](2.4)	39.0[d](3.6)
C3/%	16.4(1.2)	18.6(3.8)	16.2(2.4)	16.7(3.2)	16.9(2.8)
C4/%	34.3(3.5)	27.7(3.8)	26.6(2.6)	22.5(2.7)	22.4(3.1)
$SUVA_{254}$/[L/(m·mg)]	0.021(0.005)	0.020(0.002)	0.017(0.002)	0.019(0.003)	0.021(0.002)
E_4/E_6	8.78(0.62)	7.10(0.85)	10.32(0.98)	7.18(0.91)	7.79(0.67)
$A_{240\sim400}$	21.0(2.1)	17.9(1.5)	16.7(1.8)	15.4(2.0)	21.2(1.8)
$S_{250\sim600}$	0.0007(0.0001)	0.0006(0.0000)	0.0005(0.0001)	0.0005(0.0000)	0.0007(0.0001)

注：相同小写字母上标（C/H 和 HIX）没有显著差异，不同小写字母上标表示显著差异（$p<0.05$）。

6.1.3　土壤腐殖质转化和分解对其化学结构的影响

土壤腐殖质的转化和降解过程均受微生物活动的影响。在植物凋落物分解过程中，土壤微生物呼吸优先消耗^{13}C 的有机化合物，将富含^{13}C 的有机化合物并入土壤有机质库中（Feng，2002）。因此，$\Delta\delta^{13}$C（在本书中定义为土壤腐殖质和植物凋落物间 δ^{13}C 值的差异）通常与土壤有机质的转化和分解程度成正比（Feng，2002；Tan et al.，2013）。相关性分析结果表明，$\Delta\delta^{13}$C 与土壤胡敏酸或富里酸中的 C/H 值、HIX 和 C2 显著正相关，而与胡敏酸中的 E_4/E_6 负相关（图 6-4 和图 6-5）。土壤中腐殖质转化通常伴随着酚基官能团的初始分解产物氧化为具有含醌官能团的高聚合分子（Kawai et al.，1998；Tuor et al.，1992）。此外，土壤中腐殖质分解可能导致对醌基的优先保护（Aeschbacher et al.，2012），这可能是由于在氧化环境中醌基比其他官能团更具抗降解性（Rimmer，2006；Rimmer and Smith，2009；Rimmer and Abbott，2011）。我们的研究结果表明，土壤腐殖质的转化和降解对腐殖质的电子转移能力有积极作用。

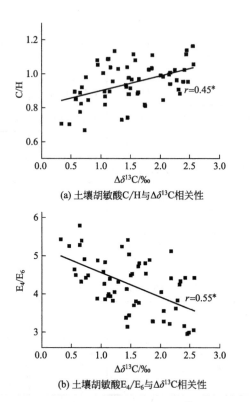

图 6-4　土壤胡敏酸 C/H 和 E4/E6 与 $\Delta\delta^{13}C$ 相关性关系

注：* 相关性在 $p<0.05$ 显著相关。

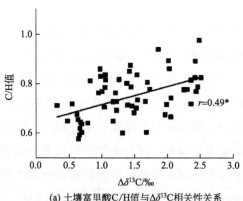

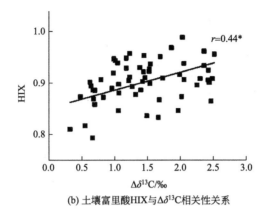

(b) 土壤富里酸HIX与$\Delta\delta^{13}$C相关性关系

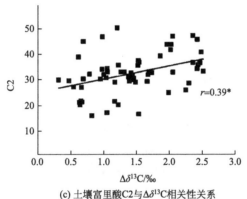

(c) 土壤富里酸C2与$\Delta\delta^{13}$C相关性关系

图 6-5　土壤富里酸 C/H 值、HIX 和 C2 与 $\Delta\delta^{13}$C 相关性关系

注：(*) 相关性在 $p<0.05$ 显著相关。

对不同农用地类型的土壤腐殖质的 $\Delta\delta^{13}$C 进行进一步比较，结果显示，水稻田土壤的腐殖质和富里酸的 $\Delta\delta^{13}$C 最高，其次分别是番茄地土壤、芹菜白菜地土壤、葡萄地土壤和杨梅地土壤（图 6-6），这表明农用地类型对土壤腐殖质的转化和分解有重要影响。基于 $\Delta\delta^{13}$C 在不同农用地类型土壤中的腐殖质和富里酸的电子转移能力一致，我们推测土壤腐殖质的转化和分解过程不同可能是由于不同农用地类型土壤腐殖质的电子转移能力不同造成的。

6.1.4　环境意义

研究农用地类型对土壤腐殖质电子转移能力的影响具有重要意义。首先，人类

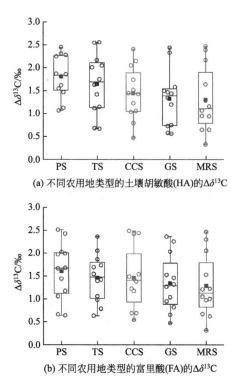

图 6-6　不同农用地类型的土壤胡敏酸（HA）和富里酸（FA）的 $\Delta\delta^{13}C$

注：不同的小写字母表示具有显著差异（$p<0.05$）。

土地利用的改变是全球变化的一个关键方面。鉴于腐殖质电子转移能力对减少土壤中有机和无机污染物的转化密切相关（Kappler and Haderlein，2003；Van der Zee and Cervantes，2009；Gu and Chen，2003；Wittbrodt and Palmer，1997），研究结果表明，从水稻田到旱地的改变可能导致土壤抗污染能力下降，这不仅威胁到土壤环境，还威胁到食品安全。其次，尽管水稻田被认为是人为甲烷的主要来源，但考虑到水稻田土壤腐殖质的电子转移能力在农用地土壤中是最大的（IPCC，2007；Tokida et al.，2011），因此水稻田土壤的腐殖质在抑制甲烷生成方面可能比其他农用地土壤发挥更大的作用（Tan et al.，2017；Tan et al.，2018）。腐殖质氧化还原过程中增加天然有机质数量和官能团含量将与 CO_2 竞争接受电子，从而减少甲烷的生成。同时，腐殖质能充当电子供体以进行反硝化作用，从而促进 N_2O 转化为 N_2（Aranda-Tamaura et al.，2007；Valenzuela et al.，2017；Martinez et al.，2013）。最后，我们的研究表明，控制环境因子（如降低农田土壤中碳氮比）以促进土壤有机质的转化和降解，以及施用有机肥来增加土壤中腐殖质的含量将成为增强腐殖质电子转移能力的有效举措，这对保障农业生态系统中土壤环境的可持续性具有重要意义。

6.2 土壤腐殖质电子转移能力对水稻田耕作年限的响应

Fe 是重要的元素，地壳中铁氧化物的含量很高。在环境中，Fe 通常以 Fe(Ⅲ) 和 Fe(Ⅱ) 两种主要的氧化还原状态存在。Fe(Ⅲ) 可以用作各种环境中微生物呼吸的末端电子受体。微生物还原 Fe(Ⅲ) 可以改变土壤和沉积物颗粒的表面特性，影响土壤和沉积物中砷（As）、铬（Cr）、铜（Cu）、铀（U）和其他无机污染物的迁移和命运。通过微生物还原 Fe(Ⅲ) 的过程产生的 Fe(Ⅱ) 可以结合以降解许多有机污染物，包括多卤代化合物、硝基芳族化合物和偶氮染料。此外，微生物中还原 Fe(Ⅲ) 还可以有效竞争电子向产甲烷菌转移，在厌氧环境中减少硝酸盐还原菌和硫酸盐还原菌，从而抑制温室气体排放。因此，环境中微生物还原 Fe(Ⅲ) 的过程与许多问题有关，从生物地球化学过程到环境修复。

细菌细胞与 Fe(Ⅲ) 矿物之间的直接接触是电子从微生物向 Fe(Ⅲ) 矿物转移的最简单策略。然而，由于 Fe(Ⅲ) 矿物的溶解性差，并且电子在细胞色素之间跳跃的距离必须小于 2.0 nm，所以除了直接接触还需要其他的机制来解释 MRF 及细胞外电子如何从细胞转移到 Fe(Ⅲ)。一些微生物可以使用 Fe(Ⅲ) 螯合剂来促进将 Fe(Ⅲ) 用作电子受体。在某些希瓦氏菌属物种和土杆菌属中，氧化还原活性纳米导线和多步电子跳跃也参与细胞外电子转移。希瓦氏菌和土杆菌属可以通过微生物或环境氧化还原活性的电子穿梭将电子转移到细胞远端的 Fe(Ⅲ) 矿物中，例如核黄素类或天然有机物质。

天然有机质是有机分子的固有异质混合物，主要来源于植物的腐烂和微生物残留，占自然环境有机物的绝大部分（MacCarthy, 2001）。天然有机质介导的矿物质包括以下两个步骤：①从微生物向天然有机质的生物电子转移；②从还原的天然有机质到铁（Ⅲ）矿物质的非生物电子转移（Piepenbrock and Kappler, 2012）。天然有机质具有电子穿梭功能的原因可能主要归因于其氧化还原活性部分，该部分在不同样品之间显示出显著差异，并且难以与环境条件形成定量关系。以前关于天然有机质电子转移能力的大多数研究都是基于从环境基质中提取的溶解相（Lovley et al., 1996; Jiang and Kappler, 2008; Tan et al.，2017a；Tan et al.，2017b）。矿物土壤中天然有机质大部分存在于固相中（Stevenson，1994；MacCarthy，2001）。最近一项基于土壤物理分馏的研究提供了证据，表明矿物土壤中固相天然有机质的电子穿梭在很大程度上取决于微生物的可及性（Tan et al.，2019）。值得注意的是，评估固相天然有机质在矿物土壤中的电子穿梭作用是通过它们与人工制备的溶解的 Fe(Ⅲ) 进行氧化还原反应。固相天然有机质的电子穿梭对原位难溶性 Fe(Ⅲ) 矿物的还原反应在矿质土壤中的作用机理仍然未知。

水稻种植在粮食安全中起着至关重要的作用，目前为全球 50％ 以上的人口提供食物（Haque et al.，2015）。因此，世界上特别是亚洲的稻田一直在不断增加。稻田的暂时缺氧状态可以通过重复的氧化还原交替驱动许多主要和微量元素的生物地球化学过程（Kögel-Knabner et al.，2010）。这些过程与铁的氧化还原循环密切相关（Borch et al.，2010）。土壤微环境会随稻田的耕作年限而变化，这种变化可能会进一步影响铁的氧化还原循环（Melton et al.，2014）。尽管已经对稻田中的矿物质进行了许多研究（Jäckel and Schnell，2000；Hori et al.，2010；Li et al.，2017），推测土壤天然有机质介导的矿物质与稻田耕作年限之间存在潜在的联系，但尚未开展相关的研究。在此，我们使用微生物还原法来评估不同耕作年限的水稻土固相天然有机物介导的矿物质。

6.2.1 原位固相天然有机质对微生物还原 Fe(Ⅲ) 矿物的影响

当将去除可提取有机质（EOM）的土壤接种 *Shewanella oneidensis* MR-1 和 *Shewanella putrefaciens* 200 并进行厌氧培养 48h，发现所有土壤样品中的 Fe(Ⅱ) 的含量与全铁含量 [Fe(总)] 的比例 [Fe(Ⅱ)/Fe(总)] 比未经处理的土壤要高（图 6-7）。该结果表明，在不存在原位固相天然有机质的情况下异化还原 Fe(Ⅲ) 的细菌可以通过直接接触机制将 Fe(Ⅲ) 矿物还原为土壤中的可溶性 Fe(Ⅱ)。由导电细菌纳米导线诱导并在微生物细胞之间产生的电子转移也可能有助于减少土壤中的 Fe(Ⅲ) 矿物（Malvankar and Lovley，2012；Nagarajan et al.，2013）。在无法获得 Fe(Ⅲ) 矿物的情况下异化 Fe(Ⅲ) 还原菌可以分泌螯合化合物，如高铁载体（Boukhalfa and Crumbliss，2002），它可以与 Fe(Ⅲ) 配合以促进 Fe(Ⅲ) 矿物的溶解，这也可能为异化 Fe(Ⅲ) 还原菌提供更多与 Fe(Ⅲ) 直接接触的机会。此外，异化 Fe(Ⅲ) 还原菌可以分泌溶解有机物，并将其用来电子穿梭以促进 Fe(Ⅲ) 矿物质的还原（Marsili et al.，2008）。

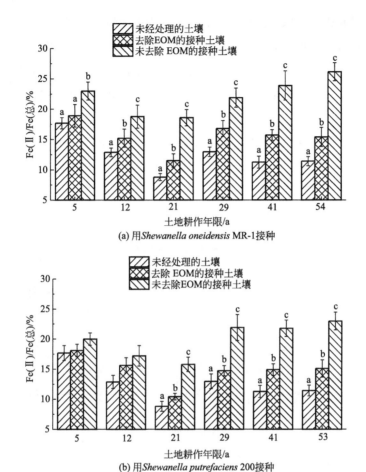

图 6-7 未经处理土壤、去除可提取有机质接种土壤和未去除可提取
有机质接种土壤中 Fe(Ⅱ)/Fe(总) 的百分比

注：Fe(Ⅱ)/Fe(总) 为土壤中 Fe(Ⅱ) 的含量与总铁含量的比例；
不同的小写字母表示具有显著差异，相同的小写字母表示没有显著差异（$p<0.05$）。

将未去除可提取有机质的土壤接种异化 Fe(Ⅲ) 还原菌并进行厌氧培养 48h 后，发现 Fe(Ⅲ) 的还原量比以相同方式培养的不含可提取有机质的土壤增加了 1~2 倍（图 6-7）。此结果表明，原位固相天然有机质在微生物还原铁(Ⅲ)矿物时起电子穿梭作用类似于溶解性可提取有机质。细菌和 Fe(Ⅲ) 矿物质之间固相天然有机质介导的电子转移可以通过电子穿梭的扩散或通过天然有机质和矿物质的网络电子跳跃来实现（Piepenbrock and Kappler，2012）。

在大多数土壤样品中,与直接电子转移、纳米导线、螯合增溶和微生物细胞间电子转移的机制相比,原位固相天然有机质作为电子转移的媒介可以为促进矿物质做出更多贡献(图 6-8)。该结果表明,固相天然有机质介导的电子转移是驱动土壤中矿物质的主要过程。土壤的致密结构以及聚集体包含的 Fe(Ⅲ) 矿物不仅可能阻碍微生物与 Fe(Ⅲ) 矿物之间的直接接触,而且会降低土壤中电子穿梭的扩散速率(Gray and Winkler, 2005)。值得注意的是,由 Fe(Ⅲ) 矿物和天然有机质形成的网络结构是从微生物到 Fe(Ⅲ) 矿物长距离电子转移的重要桥梁(Roden et al., 2010)。这可能导致由原位固相天然有机质诱导的电子转移介体的矿物质明显高于其他电子转移机制所诱导的矿物质。

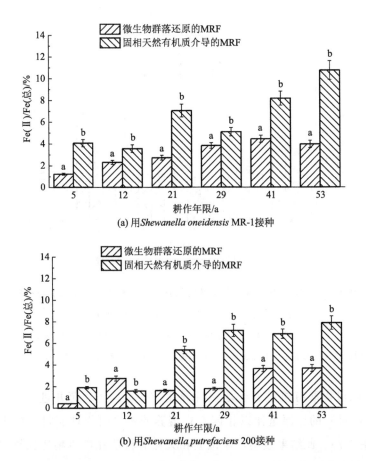

图 6-8 由微生物群落还原的 Fe(Ⅲ) 矿物质和由固相天然有机质介导的 Fe(Ⅲ) 矿物质

注:从休耕地到水稻田的不同耕作年限的比较用小写字母,不同的小写字母表示具有显著差异,相同的小写字母表示没有显著差异($p<0.05$)。

6.2.2 微生物种类对微生物还原 Fe(Ⅲ) 矿物的影响

S. oneidensis MR-1 和 S. putrefaciens 200 在还原 Fe(Ⅲ) 矿物质的能力上存在差异,在大多数土壤样品中,S. oneidensis MR-1 比 S. putrefaciens 200 在各种电子转移机理上具有更大的优势(图 6-9)。该结果表明,S. oneidensis MR-1 在土壤中具有更大的生存能力,在各种环境基质中 S. oneidensis MR-1 比 S. putrefaciens 200 的分布范围更广(DiChristina and Delong,1993;Todorova and Costello,2006;Ziemke et al.,1997)。值得注意的是,由固相天然有机质介导的微生物还原矿物质,S. oneidensis MR-1 和 S. putrefaciens 200 间存在显著相关性(图 6-10)。这表明这两个菌种可能使用相同系列的外膜蛋白介导电子从细胞内转移到细胞外,两种菌种之间电子转移绝对量的差异可能是由于它们的细胞内电子传递过程的差异所致(Myers and Myers,2000)。

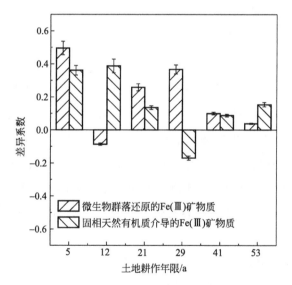

图 6-9　S. oneidensis MR-1 和 S. putrefaciens 200 还原的 Fe(Ⅲ) 矿物的差异系数
注:(*) 相关性在 $p<0.05$ 显著相关。

6.2.3 基于固相 Fe(Ⅲ) 矿物的腐殖质电子转移能力对水稻田耕作年限的响应

固相天然有机质介导的微生物还原 Fe(Ⅲ) 量随着耕作年限的增长而逐渐增加(图 6-8)。Fe(Ⅲ) 矿物的类型和组成被认为是微生物还原的关键因素。

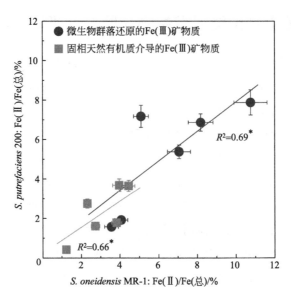

图 6-10 *S. Oneidensis* MR-1 和 *S. putrefaciens* 200 还原矿物的关系

注：(*) 相关性在 $p<0.05$ 显著相关。

随着耕作年限增加，土壤中草酸铵提取铁（Fe_o）和连二亚硫酸盐可萃取铁（Fe_d）的含量没有规律地变化（图 6-11 中曲线 a、b），而土壤中焦磷酸提取铁（Fe_p）的含量则呈逐渐增加的趋势（图 6-11 中曲线 c）。而且，固相天然有机质介导的 Fe(Ⅲ) 矿物质微生物还原量与土壤中 Fe_p 含量呈显著正相关（图 6-12）。这些结果表明，土地利用类型从休耕地变为水稻田后，固相天然有机质介导的土壤矿物质增加，可能部分归因于土壤中有机物络合态的铁含量增加，因为络合的铁可能通过与有机物紧密接触而容易接受来自还原有机物的电子（Haas and Dichristina, 2002）。

醌和对苯二酚是溶解态天然有机质中主要的氧化还原活性官能团（Kappler et al., 2004; Fimmen et al. 2007; Rakshit et al., 2009）。此外，氮和硫官能团对天然有机质的氧化还原特性也有贡献（Serudo et al., 2007; Hernández-Montoya et al., 2012）。天然有机质的氧化还原活性官能团可以通过其化学指标反映出来，包括 C/H 值、HIX、荧光组分和 E_4/E_6（Tan et al., 2017, 2018）。研究结果表明，固相天然有机质介导的微生物还原 Fe(Ⅲ) 矿物量与这些化学指标之间无显著相关性（图 6-12），表明天然有机质的氧化还原活性官能团不是控制天然有机质介导的微生物还原 Fe(Ⅲ) 矿物的主要因素。这可能是由于土壤中固相天然有机质的电子穿梭能力主要取决于其微生物的可及性而不是其化学结构（Tan et al., 2019）。值

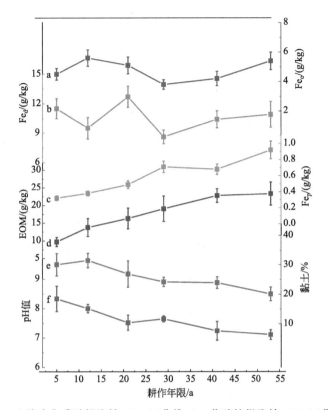

图 6-11 土壤中焦磷酸提取铁（Fe_p）(曲线 a)、草酸铵提取铁（Fe_o）(曲线 b)、
连二亚硫酸盐可萃取铁（Fe_d）(曲线 c)、可提取有机质（EOM）含量（曲线 d）
以及黏土含量（曲线 e)、pH(曲线 f) 随水稻田耕作年限的变化

得注意的是，随着耕作年限的增加，土壤中可提取有机质的含量显著增加（图6-11中曲线d），且生物还原Fe(Ⅲ)矿物质量与土壤中可提取有机质含量之间显著正相关（图6-12），这表明土地利用方式从休耕地变为水稻田后，可提取有机质含量的增加是固相天然有机质介导的土壤矿物质的增加的原因。研究结果还表明，固相天然有机质介导的矿物质可能受控于土壤中Fe(Ⅲ)矿物与天然有机质的相互作用。Fe(Ⅲ)矿物和天然有机质的络合物是土壤的核心组分。已证明土壤中天然有机质的含量对天然有机质矿物络合物的形成和微生物的还原有显著影响（Kleber et al.，2015；Haas and Dichristina，2002）。低含量的天然有机质不利于在土壤中形成天然有机质和Fe(Ⅲ)矿物的网络结构，从而导致天然有机质分子之间的距离较大，并阻碍了电子从一个天然有机质分子转移到下一个天然有机质分子（Piepenbrock and Kappler，2012）。

与Fe(Ⅲ)矿物形成紧密的物理化学键是固相天然有机质作为电子穿梭的

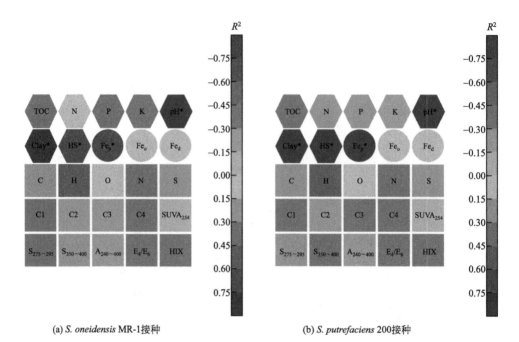

(a) *S. oneidensis* MR-1接种 (b) *S. putrefaciens* 200接种

图 6-12　固相天然有机质介导的微生物还原 Fe(Ⅲ) 矿物与土壤理化性质（六边形）、土壤中可提取态铁（圆圈）及可提取有机物化学结构的相关性（正方形）（见书后彩图 18）

注：与可提取有机物化学结构相关的矿物质量被标准化为土壤中可提取有机物的含量；颜色表示相关性的强度；（*）表示在 $p<0.05$ 显著相关。

先决条件。已有研究表明，土壤和其他环境基质中的天然有机物与黏土和铁(Ⅲ)矿物相互关联（Poulton and Raiswell, 2005；Wagai and Mayer, 2007）。Roden 等（2010）通过能量滤波转换电子显微镜（EFTEM）与电子能量损失光谱（EELS），发现在去除铁后的沉积物/Fe(Ⅲ)矿物悬浮液中，Fe(Ⅲ)纳米晶体与蒙脱石黏土矿物之间可以形成良好的聚集体。基于 EFTEM 的有机碳与铁元素图谱表明，有机碳可以直接与 Fe(Ⅲ) 纳米晶体结合，表明在土壤-氧化物悬浮液中腐殖质与 Fe(Ⅲ) 矿物在纳米级别上是相互结合的，这对于促进电子从腐殖质传递到 Fe(Ⅲ) 矿物起到十分重要的作用，进一步佐证了土壤固相腐殖质可以作为良好的电子穿梭体主要是由于其能够促进土壤腐殖质-矿物复合体网络结构的形成。

进行相关的土壤理化性质分析，以评估土地利用变化后土壤微环境变化对固相天然有机质介导的矿物质增加的贡献。土壤质地三元图显示，土地利用类型从休耕地变为水稻田后，土壤质地逐渐从黏土壤土转变为壤土（图 6-13）。随着水稻田耕作年限的增加，土壤中黏土成分的含量也相应增加[图6-11中曲线b]。这些现象以及固相天然有机质介导的Fe(Ⅲ)矿物质的微生物还原量与土壤中黏土含量显著相关（图6-12），表明从休耕地变为水稻田后，土壤质地可能是影响固相天然有机质介导的矿物质形成的重要因素。土壤中壤土的形成是有机质与矿物质相互作用的结果。尽管土壤中有些大分子腐殖质在结构上似乎要大于黏土矿物的夹层，但是土壤腐殖质可弯曲性可使其能够很好地嵌入黏土矿物的夹层中（Schnitzer et al.，1988），这种层状的硅酸盐矿物可以将土壤中分散的大分子腐殖质连接形成网络结构（Kleber and Johnson，2010），使得腐殖质分子之间的间隙不超过20 Å（1Å=10^{-10}m）（Piepenbrock and Kappler，2012），这为电子在相邻腐殖质分子之间的跃迁提供了可能。黏土质地的土壤易于团聚，不利于形成良好的土壤结构，使得土壤中存在较多的缺失有机质的斑块（Peth et al.，2008），阻断了电子在腐殖质分子间的跳跃，从而大大降低了土壤固相腐殖质的电子转移能力。另外，黏土质地土壤中的Fe(Ⅲ)矿物的微生物可利用的表面积相对较低，这也可能限制了微生物对Fe(Ⅲ)的还原（Roden and Zachara，1996）。

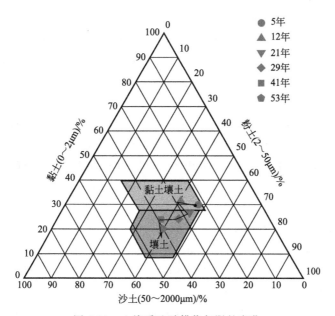

图6-13　土壤质地随耕作年限的变化

土壤铁矿物质，包括赤铁矿、黄铁矿、磁铁矿、菱铁矿和针铁矿，在不同的pH条件下具有不同的氧化还原电位（Johnson et al.，2012；Majzlan，2012）。从休耕地到水稻田，土壤pH值随着耕作年限的增加而呈下降趋势（图6-11中曲线 f)，这导致固相天然有机质介导的微生物还原的 Fe(Ⅲ) 矿物质的量与土壤中可提取有机质含量之间显著负相关（图 6-12）。土壤 pH 的降低可能导致土壤铁矿物质的氧化还原电位降低，从而降低铁矿物质接受电子的能力，这可能是土地利用类型从休耕地变为水稻田后土壤中固相天然有机质介导的矿物质增加的另一个原因。

参考文献

Aeschbacher M，Sander M，Schwarzenbach R P，2010. Novel electrochemical approach to assess the redox properties of humic substances. Environmental Science & Technology, 44 (1)：87-93.

Aeschbacher M，Vergari D，Schwarzenbach R P，et al，2011. Electrochemical analysis of proton and electron transfer equilibria of the reducible moieties in humic acids. Environmental Science & Technology, 45 (19)：8385-8394.

Aeschbacher M，Graf C，Schwarzenbach R P，et al，2012. Antioxidant properties of humic substances. Environmental Science & Technology, 46 (9)，4916-4925.

Aiken G R，McKnight D M，Wershaw R. L.，et al，1985. Humic substances in soil, Sediment and water：Geochemistry, isolation and characterization. New York：John Wiley & Sons.

Aranda-Tamaura C，Estrada-Alvarado M I，Texier A C，et al，2007. Effects of different quinoid redox mediators on the removal of sulphide and nitrate via denitrification. Chemosphere，69 (11)：1722-1727.

Bauer I，Kappler A，2009. Rates and extent of reduction of Fe(Ⅲ) compounds and O_2 by humic substances. Environmental Science & Technology, 43 (13)：4902-4908.

Benz M，Schink B，Brune A，1998. Humic acid reduction by *Propionibacterium freudenreichii* and other fermenting bacteria. Applied and Environmental Microbiology, 64 (11)：4507-4512.

Borch T，Kretzschmar R，Kappler A，et al，2010. Biogeochemical redox processes and their impacton contaminant dynamics. Environmental Science & Technology, 44 (1)：15-23.

Boukhalfa H，Crumbliss A L，2002. Chemical aspects of siderophore mediated iron transport. Biometals, 15 (4)：325-339.

Cervantes F J，de Bok F A M，Tuan D D，et al，2002. Reduction of humic substances by halorespiring, sulphate-reducing and methanogenic microorganisms. Environmental Microbiology, 4 (1)：51-57.

Climate change，2007. the physical science basis, IPCC, Cambridge University Press, Cambridge, United Kingdom, New York, NY, USA.

DiChristina T J，Delong E F，1993. Design and application of rRNA-targeted oligonucleotide probes for the dissimilatory iron-and manganese-reducing bacterium Shewanella putrefaciens. Applied and Environmental Microbiology, 59 (12): 4152-4160.

Dunnivant F M, Schwarzenbach R P, Macalady D L, 1992. Reduction of substituted nitrobenzenes in aqueous solutions containing natural organic matter. Environmental Science & Technology, 26 (11): 2133-2141.

Einsiedl F, Mayer B, Schafer T, 2008. Evidence for incorporation of H_2S in groundwater fulvic acids from stable isotope ratios and sulfur K-edge X-ray absorption near edge structure spectroscopy. Environmental Science & Technology, 42 (7): 2439-2444.

Feng X, 2002. A theoretical analysis of carbon isotope evolution of decomposing plant litters and soil organic matter. Global Biogeochemical Cycles, 16 (4): 1119-1130.

Feng X, Simpson A J, Wilson K P, et al, 2008. Increased cuticular carbon sequestration and ligin oxidation in response to soil warming. Nature Geoscience, 1 (12): 836-839.

Fimmen R L, Cory R M, Chin Y P, et al, 2007. Probing the oxidation-reduction properties of terrestrially and microbially derived dissolved organic matter. Geochimica et Cosmochimica Acta, 71 (12): 3003-3015.

Gray H B, Winkler J R, 2005. Long-range electron transfer. Proceedings of the National Academy of Sciences, 102 (10): 3534-3539.

Gu B H, Chen J, 2003. Enhanced microbial reduction of Cr(Ⅵ) and U(Ⅵ) by different natural organic matter fractions. Geochimica et Cosmochimica Acta, 67 (19): 3575-3582.

Haas J R, Dichristina T J, 2002. Effects of Fe(Ⅲ) chemical specation on dissimilatory Fe(Ⅲ) reduction by Shewanella putrefaciens. Environmental Science & Technology, 36 (3): 373-380.

Haque M M, Kim S Y, Ali M A, et al, 2015. Contribution of greenhouse gas emissions during cropping and fallow seasons on total global warming potential in mono-rice paddy soils. Plant Soil, 387: 251-264.

Hernández-Montoya V, Alvarez L H, Montes-Morán M A, et al, 2012. Reduction of quinone and non-quinone redox functional groups in different humic acid samples by Geobacter sulfurreducens. Geoderma, 183-184: 25-31.

Herzsprung P, Tuempling W V, Hertkorn N, et al. Variations of DOM Quality in Inflows of a Drinking Water Reservoir: Linking of van Krevelen Diagrams with EEMF Spectra by Rank Correlation [J]. Environmental Science & Technology, 2012, 46 (10): 5511-5518.

Hori T, Müller A, Igarashi Y, et al, 2010. Identification of iron-reducing microorganisms in anoxic rice paddy soil by ^{13}C-acetate probing. Isme Journal, 4 (2): 267-278.

IPCC, 2007. Climate change 2007: the physical science basis. Cambridge University Press, Cambridge, United Kingdom, New York, NY, USA.

Jäckel U, Schnell S, 2000. Role of microbial iron reduction in paddy soil//van Ham J, Baede A P M, Meyer L A, Ybema R. Non-CO_2 Greenhouse Gases: Scientific Understanding, Control and Implementation. Springer, Dordrecht, 143-144.

Jiang J, Kappler A, 2008. Kinetics of microbial and chemical reduction of humic substances: Implications for electron shuttling. Environmental Science & Technology, 42 (12): 3563-3569.

Johnson D B, Kanao T, Hedrich S, 2012. Redox transformations of iron at extremely low pH: Fundamental and applied aspects. Frontiers in Microbiology, 3 (96): 96.

Kappler A, Benz M, Schink B, et al, 2004. Electron shuttling via humic acids in microbial iron (Ⅲ) reduction in a fresh water sediment. FEMS Microbiology Ecology, 47: 85-92.

Kappler A, Haderlein S B, 2003. Natural organic matter as reductant for chlorinated aliphatic pollutants. Environmental Science & Technology, 37 (12): 2714-2719.

Kawai S, Umezawa T, Higuchi T, 1998. Degradation mechanisms of phenolic b-1 lignin substructure model compounds by laccase of coriolus versicolor. Archives of Biochemis and Biophysics, 262 (1): 99-110.

Kleber M, Johnson M G, 2010. Advances in understanding the molecular structure of soil organic matter: Implications for interactions in the environment. Advances in Agronomy, 106: 77-142.

Kleber M, Eusterhues K, Keiluweit M, et al, 2015. Mineral-organic associations: Formation, properties, and relevance in soil environments. Advances in Agronomy, 130: 1-140.

Kögel-Knabner I, Amelung W, Cao Z H, et al, 2010. Biogeochemistry of paddy soils. Geoderma. 157 (1-2): 1-14.

Li L, Qu Z, Jia R, et al, 2017. Excessive input of phosphorus significantly affects microbial Fe(Ⅲ) reduction in flooded paddy soils by changing the abundances and community structures of *Clostridium* and *Geobacteraceae*. Science of the Total Environment, 607-608: 982-991.

Lovley D R, Coates J D, BluntHarris E L, et al, 1996. Humic substances as electron acceptors for microbial respiration. Nature, 382 (6590): 445-448.

MacCarthy P, 2001. The principles of humic substances. Soil Science, 166 (11): 738-751.

Majzlan J, 2012. Minerals and aqueous species of iron and manganese as reactants and products of microbial metal respiration. //Gescher J, Kappler A. Microbial Metal Respiration. Springer, Heidelberg, 1-28.

Malvankar N S, Lovley D R, 2012. Microbial nanowires: A new paradigm for biological electron transfer and bioelectronics. Chemsuschem, 5 (6): 1039-1046.

Marsili E, Baron D B, Shikhare I D, et al, 2008. *Shewanella* secretes flavins that mediate extracellular electron transfer. Proceedings of the National Academy of Sciences, 105 (10): 3968-3973.

Martinez C M, Alvarez L H, Celis L B, et al, 2013. Humus-reducing microorganisms and their valuable contribution in environmental processes. Applied and Environmental Microbiology, 97 (24): 10293-10308.

Melton E D, Swanner E D, Behrens S, et al, 2014. The interplay of microbially mediated and abiotic reactions in the biogeochemical Fe cycle. Nature Reviews Microbiology, 12 (12): 797-808.

Myers J M, Myers C R, 2000. Role of the tetraheme cytochrome CymA in anaerobic electron transport in cells of *Shewanella putrefaciens* MR-1 with normal levels of menaquinone. Journal of Bacteriology, 182 (1): 67-75.

Nagarajan H, Embree M, Rotaru A E, et al, 2013. Characterization and modelling of interspecies electron transfer mechanisms and microbial community dynamics of a syntrophic association. Nature Communications, 4 (7): 2809.

Newman D K, Kolter R A, 2000. Role of excreted quinones in extracellular electron transfer. Nature, 405 (6782): 94-97.

Ohno T, 2002. Fluorescence inner-filtering correction for determining the humification index of dissolved organic matter. Environmental Science & Technology, 36 (19): 742-746.

Peth S, Horn R, Beckmann F, et al, 2008. Three-dimensional quantification of intra-aggregate pore-space features using synchrotron-radiation-based microtomography. Soil Science Society of America Journal, 72 (4): 897-907.

Piepenbrock A, Kappler A, 2012. Humic substances and extracellular electron transfer. //Gescher J, Kappler A. Microbial Metal Respiration. Springer, Heidelberg, 107-128.

Pisani O, Hills K M, Courtier-Murias D, et al, 2014. Accumulation of aliphatic compounds in soil with increasing mean annual temperature. Organic Geochemistry, 76: 118-127.

Pisani O, Frey S D, Simpson A J, et al, 2015. Soil warming and nitrogen deposition alter soil organic matter composition at the molecular-level. Biogeochemistry, 123 (3): 391-409.

Poulton S W, Raiswell R, 2005. Chemical and physical characteristics of iron oxides in riverine and glacial meltwater sediments. Chemical Geology, 218 (3-4): 203-221.

Rakshit S, Uchimiya M, Sposito G, 2009. Iron (Ⅲ) bioreduction in soil in the presence of added humic substances. Soil Science Society America Joural, 73: 65-71.

Rimmer D L, 2006. Free radicals, antioxidants, and soil organic matter recalcitrance. European Journal of Soil Science, 57 (2): 91-94.

Rimmer D L, Smith A M, 2009. Antioxidants in soil organic matter and in associated plant materials. European Journal of Soil Science, 60 (2): 170-175.

Rimmer D L, Abbott G D, 2011. Phenolic compounds in NaOH extracts of UK soils and their contribution to antioxidant capacity. European Journal of Soil Science, 62 (2): 285-294.

Roden E E, Zachara J M, 1996. Microbial reduction of crystalline iron (Ⅲ) oxides: Influence of oxide surface area and potential for cell growth. Environmental ence and Technology, 30 (5): 1618-1628.

Roden E E, Kappler A, Bauer I, et al, 2010. Extracellular electron transfer through microbial reduction of solid-phase humic substances. Nature Geoscience, 3 (6): 417-421.

Schnitzer M, Ripmeester J A, Kodama H, 1988. Characterization of the organic-matter associated with a soil clay. Soil Science, 145 (6): 448-454.

Scott D T, McKnight D M, Blunt-Harris E L, et al, 1998. Quinone moieties act as electron acceptors in the reduction of humic substances by humics-reducing microorganisms. Environmental Science & Technology, 32 (19): 2984-2989.

Serudo R L, de Oliveira L C, Rocha J C, et al, 2007. Reduction capability of soil humic substances from the Rio Negro basin, Brazil, towards Hg(II) studied by a multimethod approach and principal component analysis (PCA). Geoderma, 138: 229-236.

Shaver G R, Canadell J, Chapin F S, et al, 2000. Global warming and terrestrial ecosystems: A conceptual framework for analysis. BioScience, 50 (10): 871-882.

Stevenson F J, 1994. Humus chemistry: genesis, composition, reactions. John Wiley & Sons, New York.

Struyk Z, Sposito G, 2001. Redox properties of standard humic acids. Geoderma, 102 (3-4): 329-346.

Tan W, Zhang Y, Xi B, et al, 2018. Discrepant responses of the electron transfer capacity of soil humic substances to irrigations with wastewaters from different sources. Science of The Total Environment, 610-611: 333-341.

Tan W, Xi B, Wang G, et al, 2017. Increased electron-accepting and decreased electron-donating capacities of soil humic substances in response to increasing temperature. Environmental Science & Technology, 51 (6): 3176-3186.

Tan W, Zhou L, Liu K, 2013. Soil aggregate fraction-based ^{14}C analysis and its application in the study of soil organic carbon turnover under forests of different ages. Chinese Science Bulletin, 58 (16): 1936-1947.

Tan W, Xi B, Wang G, et al, 2019. Microbial-accessibility-dependent electron shuttling of in situ solid-phase organic matter in soils. Geoderma, 338: 1-4.

Todorova S G, Costello A M, 2006. Design of *Shewanella* - specific 16S rRNA primers and application to analysis of *Shewanella* in a minerotrophic wetland. Environmental Microbiology, 8 (3): 426-432.

Tokida T, Adachi M, Cheng W, et al, 2011. Methane and soil CO_2 production from current-season photosynthates in a rice paddy exposed to elevated CO_2 concentration and soil temperature. Global Change Biology, 17 (11): 3327-3337.

Tratnyek P G, Macalady D L, 1989. Abiotic reduction of nitro aromatic pesticides in anaerobic laboratory systems. Journal of Agricultural and Food Chemistry, 37 (1): 248-254.

Tuor U, Wariishi H, Schoemaker H E, et al, 1992. Oxidation of phenolic arylglycerol b-aryl ether lignin model compounds by manganese peroxidase from Phenerochaete chrysosporium: Oxidative cleavage of an a-carbonyl model compound. Biochemistry, 31 (21): 4986-4995.

Wagai R, Mayer L M, 2007. Sorptive stabilization of organic matter in soils by hydrous iron oxides. Geochimica Et Cosmochimica Acta, 71 (1): 25-35.

Wittbrodt P R, Palmer C D, 1997. Reduction of Cr(Ⅵ) by soil humic acids. European Journal of Soil Science, 48 (1): 151-162.

Valenzuela E I, Prietodavó A, Lópezlozano N E, et al, 2017. Anaerobic methane oxidation driven by microbial reduction of natural organic matter in a tropical wetland. Applied and Environmental Microbiology, 83 (11).

Van der Zee F R, Cervantes F J, 2009. Impact and application of electron shuttles on the redox (bio) transformation of contaminants: A review. Biotechnology Advances, 27 (3): 256-277.

Ziemke F, Brettar I, Höfle M G, 1997. Stability and diversity of the genetic structure of a *Shewanella putrefaciens* population in the water column of the central Baltic. Aquatic Microbial Ecology, 13: 63-74.

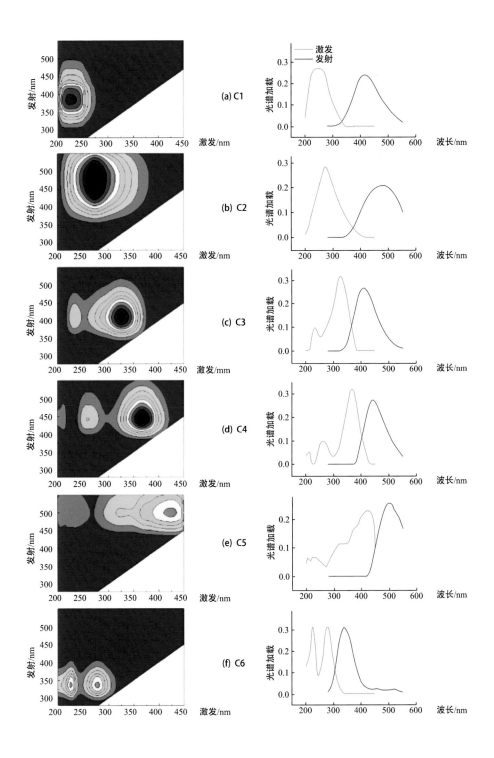

彩图1 PARAFAC分析确定的六种组分验证模型的激发和发射负载

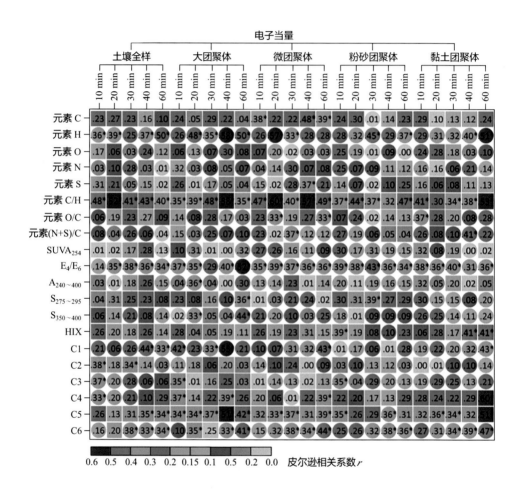

彩图2 不同微生物转移到溶解性HS的电子当量与化学结构参数之间的相关性

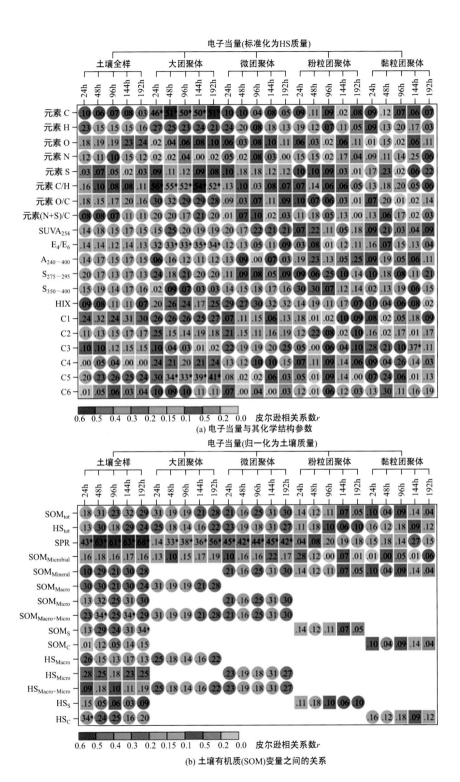

彩图3 不同微生物经一定还原时间后转移到固相HS的电子当量与其化学结构参数

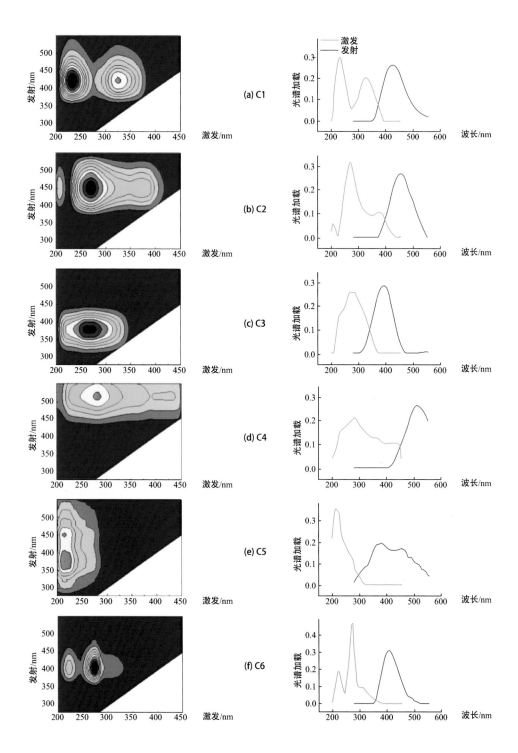

彩图4 PARAFAC分析确定的六种组分验证模型的激发和发射负载

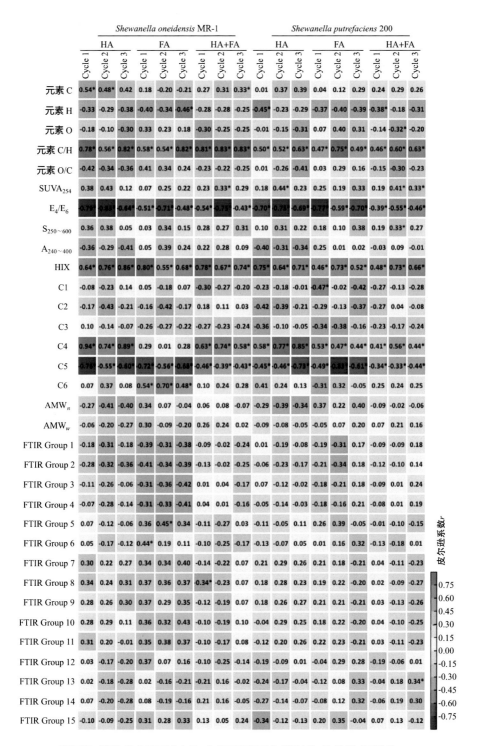

彩图5 微生物还原能力和表征腐殖质化学结构的参数相关性

注：HA+FA表示统计分析中胡敏酸和富里酸的集合数据集；红外官能团1~15表示通过红外光谱分析出的功能性基团；颜色和数字表示相关性的正负性和强度；(*)表示显著相关($p<0.05$)；Cycle1、Cycle2、Cycle3分别表示第一次循环、第二次循环、第三次循环。

	Shewanella oneidensis MR-1									Shewanella putrefaciens 200								
	HA			FA			HA+FA			HA			FA			HA+FA		
	Cycle 1	Cycle 2	Cycle 3	Cycle 1	Cycle 2	Cycle 3	Cycle 1	Cycle 2	Cycle 3	Cycle 1	Cycle 2	Cycle 3	Cycle 1	Cycle 2	Cycle 3	Cycle 1	Cycle 2	Cycle 3
元素 C	0.31	0.43	0.36	-0.32	-0.30	-0.36	-0.25	-0.30	-0.21	0.23	0.41	0.42	-0.34	-0.34	-0.40	-0.27	-0.29	-0.31
元素 H	0.10	0.33	0.23	-0.18	-0.14	-0.15	0.15	0.19	0.18	0.17	0.38	0.36	0.04	0.06	0.15	0.26	0.30	0.36*
元素 O	-0.15	-0.05	-0.10	-0.30	-0.38	-0.30	0.26	0.24	0.28	-0.19	-0.01	-0.01	-0.44*	-0.30	-0.30	0.23	0.31	0.29
元素 C/H	0.19	-0.10	-0.04	-0.11	-0.11	0.14	-0.23	-0.19	-0.24	-0.05	-0.15	-0.14	-0.25	-0.27	-0.32	-0.27	-0.27	-0.26
元素 O/C	-0.31	-0.23	-0.27	-0.23	-0.17	-0.20	0.30	0.19	0.19	-0.21	-0.26	-0.19	-0.36	-0.39	-0.37	0.29	0.31	0.28
SUVA$_{254}$	0.35	0.24	0.16	0.12	0.29	0.31	-0.22	-0.12	-0.23	0.17	0.31	0.27	0.07	-0.13	0.01	-0.29	-0.27	-0.30
E$_4$/E$_6$	-0.34	-0.12	-0.10	0.28	0.05	0.03	0.24	0.14	0.27	-0.12	-0.04	-0.01	0.21	0.47*	0.40	0.31	0.27	0.22
S$_{250\sim600}$	0.33	0.28	0.18	0.18	0.36	0.40	-0.25	-0.26	-0.21	0.12	0.31	0.31	0.08	-0.09	0.05	-0.28	-0.30	-0.23
A$_{240\sim400}$	0.11	0.26	0.27	0.40	0.41	0.40	-0.20	-0.35*	-0.30	-0.05	0.17	0.23	0.23	0.18	0.21	-0.25	-0.16	-0.16
HIX	0.39	0.42	0.36	-0.33	-0.33	-0.43	-0.27	-0.29	-0.31	0.45*	0.39	0.39	-0.37	-0.42	-0.35	-0.31	-0.28	-0.30
C1	0.41	0.40	0.33	0.09	-0.09	-0.13	0.24	0.29	0.26	0.36	0.43	0.41	-0.07	0.19	0.06	0.27	0.23	0.25
C2	0.06	0.11	0.18	-0.04	-0.17	-0.23	-0.30	-0.23	-0.31	0.02	0.01	0.05	-0.18	0.14	-0.03	-0.28	-0.24	-0.34*
C3	-0.35	-0.36	-0.39	0.34	0.24	0.25	0.27	0.17	0.24	-0.30	-0.33	-0.35	0.34	0.39	0.42	0.31	0.25	0.24
C4	0.24	-0.06	-0.02	-0.16	-0.34	-0.33	-0.23	-0.33*	-0.23	0.01	0.01	-0.01	-0.32	-0.08	-0.24	-0.27	-0.22	-0.24
C5	-0.18	0.09	0.08	0.40	0.33	0.37	-0.23	-0.03	-0.12	0.09	0.07	0.05	0.36	0.39	0.34	-0.04	-0.04	-0.04
C6	0.38	0.40	0.41	-0.30	-0.08	-0.07	0.14	0.18	0.24	0.40	0.34	0.34	-0.15	-0.47*	-0.35	0.19	-0.03	0.04
AMW$_n$	-0.73*	-0.80*	-0.80*	-0.74*	-0.74*	-0.66*	-0.82*	-0.67*	-0.83*	-0.75*	-0.80*	-0.80*	-0.49*	-0.70*	-0.70*	-0.70*	-0.75*	-0.69*
AMW$_w$	-0.80*	-0.80*	-0.77*	-0.80*	-0.65*	-0.78*	-0.80*	-0.70*	-0.80*	-0.72*	-0.80*	-0.80*	-0.59*	-0.67*	-0.68*	-0.80*	-0.80*	-0.80*
FTIR Group 1	-0.30	-0.36	-0.34	-0.09	0.03	0.03	-0.30	-0.22	-0.21	-0.36	-0.30	-0.32	0.03	-0.04	0.07	-0.22	-0.29	-0.18
FTIR Group 2	-0.37	-0.42	-0.41	0.07	0.10	0.14	-0.29	-0.19	-0.19	-0.38	-0.31	-0.27	0.10	0.02	0.16	-0.25	-0.26	-0.23
FTIR Group 3	-0.23	-0.31	-0.21	-0.24	-0.13	-0.11	-0.27	-0.31	-0.29	-0.24	-0.53*	-0.42	-0.10	-0.13	-0.06	-0.26	-0.27	-0.28
FTIR Group 4	-0.25	-0.30	-0.24	-0.26	-0.15	-0.14	-0.30	-0.29	-0.27	-0.36	-0.31	-0.37	-0.13	-0.10	-0.10	-0.32*	-0.22	-0.31
FTIR Group 5	-0.09	-0.17	-0.12	-0.14	-0.11	-0.10	0.25	0.29	0.26	-0.27	-0.36	-0.28	-0.11	-0.06	-0.16	0.29	0.26	0.23
FTIR Group 6	-0.16	-0.22	-0.22	-0.32	-0.35	-0.29	0.15	0.01	0.08	-0.34	-0.36	-0.37	-0.25	-0.31	0.47*	0.07	-0.07	-0.03
FTIR Group 7	-0.14	-0.40	-0.32	0.35	0.17	0.20	0.25	0.21	0.21	-0.43	-0.40	-0.37	0.22	0.18	0.13	0.19	0.19	0.21
FTIR Group 8	-0.03	-0.22	-0.27	0.27	0.22	0.18	0.31	0.18	0.19	-0.27	-0.28	-0.27	0.17	0.22	0.13	0.25	0.18	0.16
FTIR Group 9	0.03	-0.14	-0.16	0.36	0.20	0.23	0.28	0.23	0.24	-0.21	-0.22	-0.20	0.19	0.22	0.10	0.27	0.23	0.22
FTIR Group 10	-0.03	-0.16	-0.16	0.31	0.16	0.18	0.24	0.17	0.21	-0.24	-0.20	-0.23	0.16	0.21	0.12	0.21	0.20	0.20
FTIR Group 11	-0.12	-0.23	-0.24	0.28	0.19	0.18	0.23	0.20	0.23	-0.31	-0.25	-0.26	0.15	0.20	0.10	0.22	0.16	0.17
FTIR Group 12	-0.10	-0.12	-0.13	-0.36	-0.33	-0.31	-0.09	-0.08	-0.10	-0.29	-0.34	-0.30	-0.29	-0.37	-0.28	-0.05	-0.09	-0.09
FTIR Group 13	-0.11	-0.12	-0.14	-0.40	-0.41	-0.28	-0.23	-0.18	-0.21	-0.34	-0.27	-0.27	-0.37	-0.27	-0.36	-0.24	-0.20	-0.16
FTIR Group 14	-0.20	-0.18	-0.21	-0.18	-0.21	-0.37	-0.19	-0.20	-0.21	-0.30	-0.31	-0.29	-0.21	-0.18	-0.17	-0.31	-0.18	-0.17
FTIR Group 15	0.36	0.39	0.26	-0.11	-0.24	-0.22	0.21	0.08	0.11	0.29	0.37	0.35	-0.25	-0.16	-0.31	0.07	0.11	0.06

彩图6 还原能力恢复程度和表征腐殖质化学结构的参数相关性

注：HA+FA表示统计分析中胡敏酸和富里酸的集合数据；红外官能团1~15表示通过红外光谱分析出的功能性基团；颜色和数字表示相关性的正负性和强度；(*)表示在$p<0.05$显著相关。

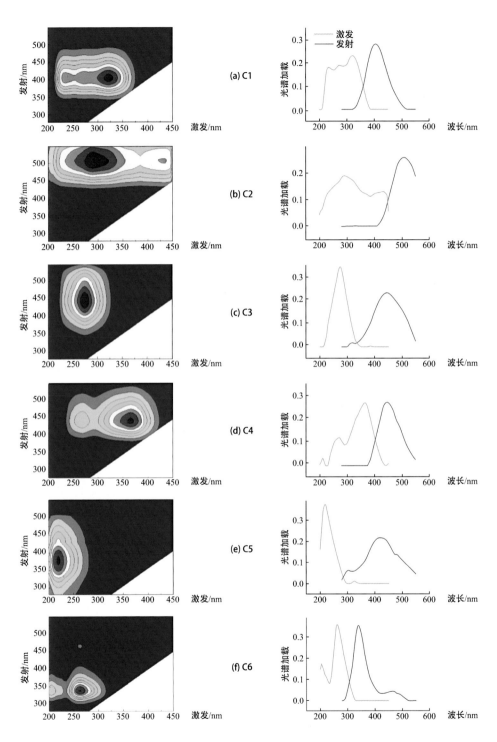

彩图7　PARAFAC分析确定的六种组分验证模型的激发和发射负载

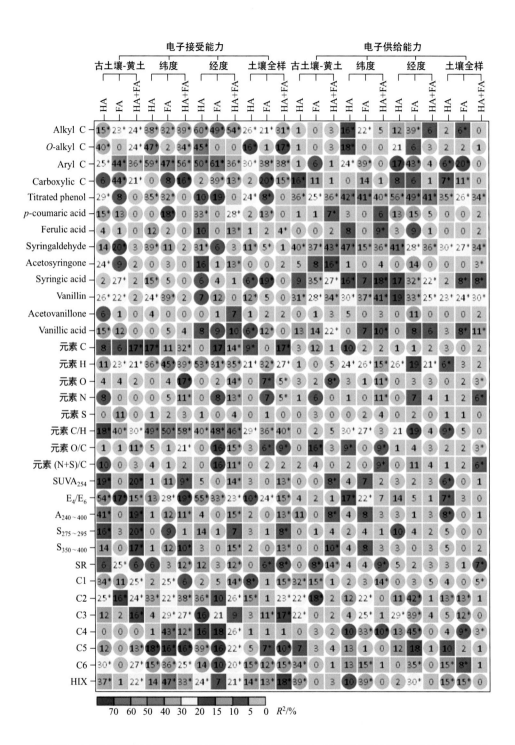

彩图8 土壤腐殖质的电子供给能力和电子接受能力与其物理化学结构的相关系数

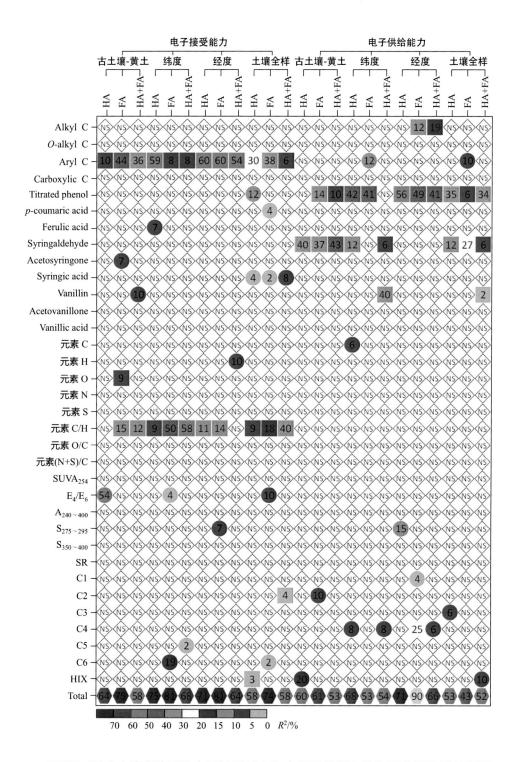

彩图9 影响土壤腐殖质的电子接受能力和电子供给能力的物理化学性质的指标

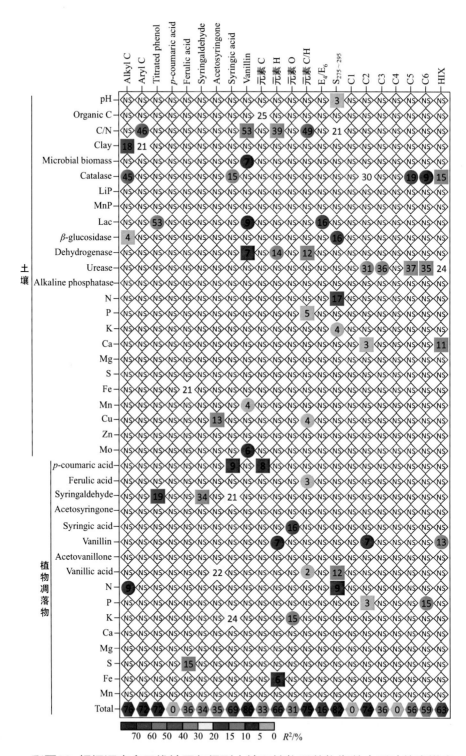

彩图10 根据逐步多元线性回归得到土壤和植物凋落物指数来反映纬度梯度土壤胡敏酸物理化学性质的指标

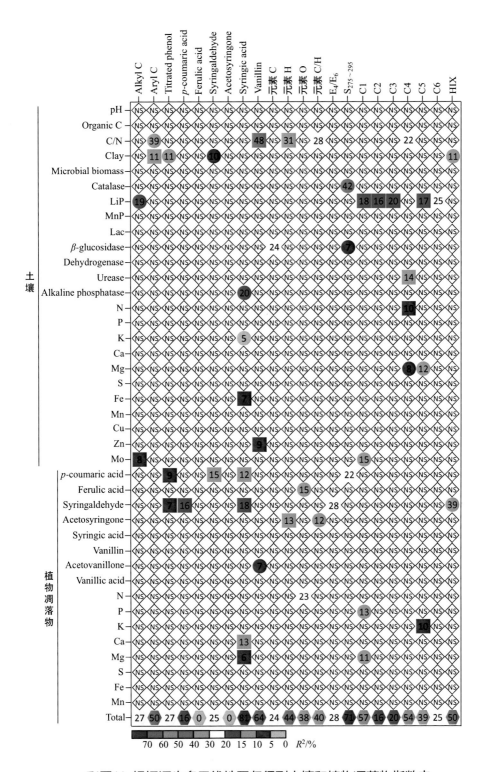

彩图11 根据逐步多元线性回归得到土壤和植物凋落物指数来反映纬度梯度土壤富里酸物理化学性质的指标

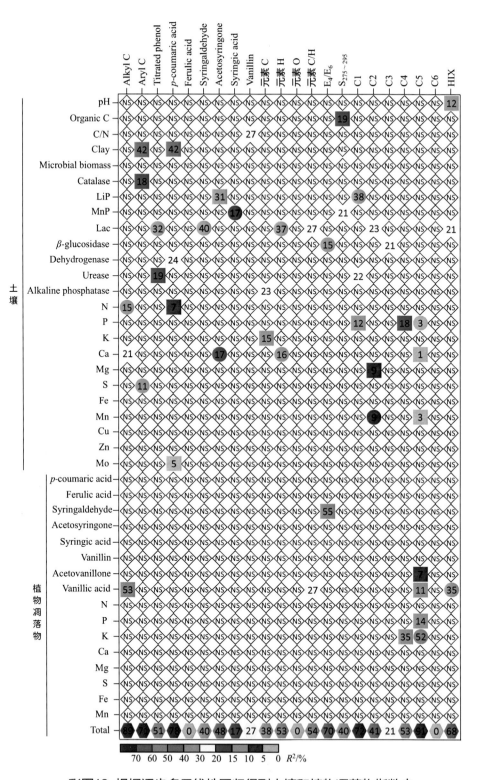

彩图12 根据逐步多元线性回归得到土壤和植物凋落物指数来反映海拔梯度土壤胡敏酸物理化学性质的指标

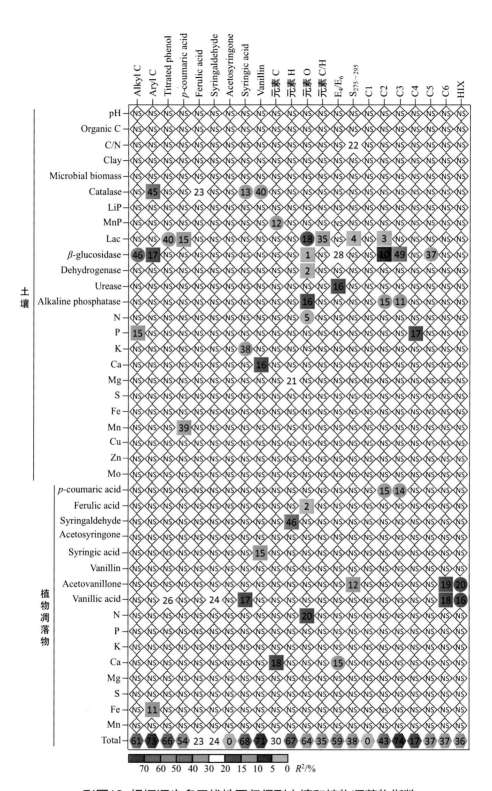

彩图13 根据逐步多元线性回归得到土壤和植物凋落物指数来反映海拔梯度土壤富力酸物理化学性质的指标

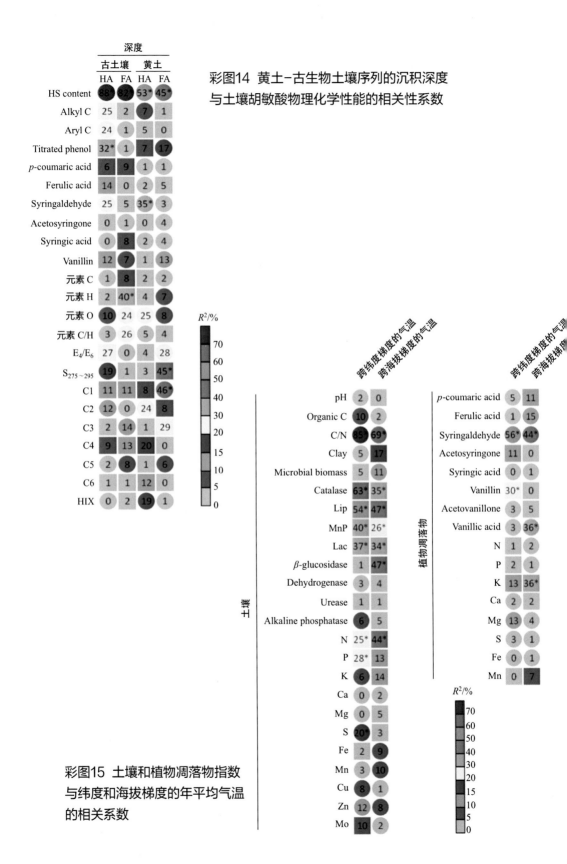

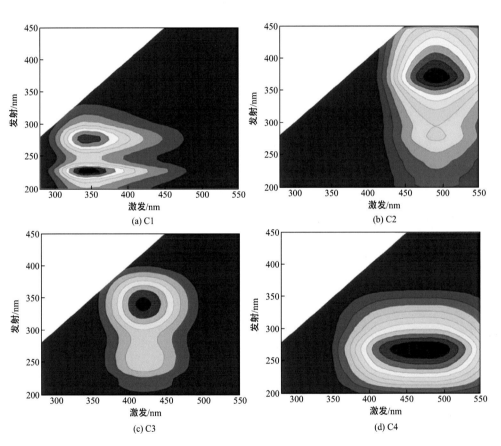

彩图16 使用平行因子分析确定的荧光成分

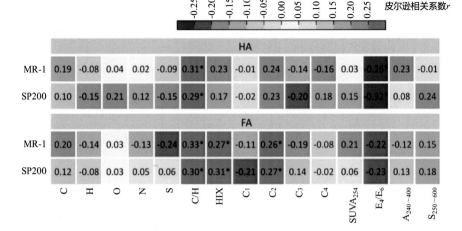

彩图17 S. oneidensis MR-1和S. putrefaciens 200接种下
电子转移能力与胡敏酸和富里酸的化学结构因子的相关性

注：颜色和数字表示相关强度，显著相关(*)在$p<0.05$水平上进行评估。

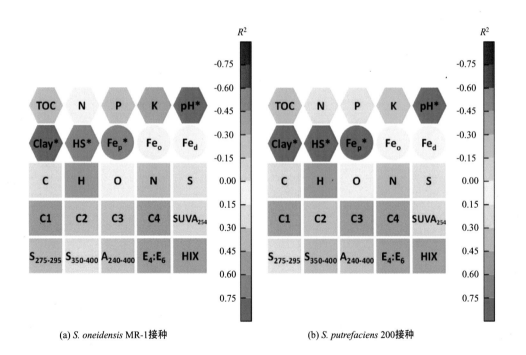

(a) S. oneidensis MR-1接种 (b) S. putrefaciens 200接种

彩图18 固相天然有机质介导的微生物还原Fe（Ⅲ）矿物与土壤理化性质（六边形）、
土壤中可提取态铁（圆圈）及可提取有机物化学结构的相关性（正方形）

注：与可提取有机物化学结构相关的矿物质量被标准化为土壤中可提取有机物的含量；
颜色表示相关性的强度；(*)相关性在$p<0.05$显著相关。